Hydrogen Production and Storage

Topical Conference at the 2010 AIChE Annual Meeting

Salt Lake City, Utah, USA
7-12 November 2010

ISBN: 978-1-61782-139-4

Printed from e-media with permission by:

Curran Associates, Inc.
57 Morehouse Lane
Red Hook, NY 12571

CURRAN ASSOCIATES INC.
proceedings
.com

Some format issues inherent in the e-media version may also appear in this print version.

Printed by Curran Associates, Inc. (2011)

AIChE
3 Park Avenue
New York, NY 10016-5991

Phone: (203) 702-7660
Fax: (203) 775-5177

www.aiche.org

Additional copies of this publication are available from:

Curran Associates, Inc.
57 Morehouse Lane
Red Hook, NY 12571 USA
Phone: 845-758-0400
Fax: 845-758-2634
Email: curran@proceedings.com
Web: www.proceedings.com

TABLE OF CONTENTS

(26a) Compact Hydrogen Storage with Extended Dormancy In Cryogenic Pressure Vessels

salvador M. Aceves, Engineering, Lawrence Livermore National Laboratory, Livermore, CA

LLNL is developing cryogenic capable pressure vessels with thermal endurance 5-10 times greater than conventional liquid hydrogen (LH2) tanks that can eliminate evaporative losses in routine usage of (L) H2 automobiles. Cryogenic pressure vessels can be fueled with ambient temperature compressed gaseous hydrogen (CGH2), LH2 or cryogenic hydrogen at elevated supercritical pressure (cryo-compressed hydrogen, CcH2). When filled with LH2 or CcH2, these vessels contain 2-3 times more fuel than conventional ambient temperature compressed H2 vessels. LLNL has demonstrated fueling with LH2 onboard two vehicles. The generation 2 vessel, installed onboard an H2-powered Toyota Prius and fueled with LH2 demonstrated the longest unrefueled driving distance and the longest cryogenic H2 hold time without evaporative losses. A third generation vessel has been installed, reducing weight and volume by minimizing insulation thickness while still providing acceptable thermal endurance.

See more of this Session: Hydrogen Storage System Engineering and Applications: Infrastructure, Delivery Systems and Demonstration Systems
See more of this Group/Topical: Topical 8: Hydrogen Production and Storage

(26b) Evaluation of Sorption System for Hydrogen Storage

Richard Chahine, Hydrogen Research Institute, Trois-Rivieres, QC, Canada

Significant R&D efforts are being pursued globally to improve the efficiency and economics of onboard hydrogen storage systems for mobile applications. The sorption of hydrogen on or in a solid substrate, relying either on physisorption or chemisorption or even a combination of both offers a potential alternative means of storing hydrogen. Hydrogen storage via physisorption on porous materials, such as activated carbon, carbon nanostructures and metal organic frameworks (MOFs) is particularly enticing due to its inherent reversibility and cyclability. Moreover, these materials exhibit fast kinetics and operate at relatively low storage pressures. However, hydrogen physisorption at ambient temperatures on currently available porous materials leads to unacceptably low hydrogen storage densities. Acceptable densities are only attainable at cryogenic temperatures.

We will present the results of our evaluation of an adsorption-based hydrogen storage tank for vehicular application. We will discuss the net storage capacity of the system over wide temperature and pressure ranges, the thermal requirements to charge and discharge hydrogen at the desired flow rates, and dormancy.

See more of this Session: Hydrogen Storage System Engineering and Applications: Infrastructure, Delivery Systems and Demonstration Systems
See more of this Group/Topical: Topical 8: Hydrogen Production and Storage

(26c) JHFC (Japan Hydrogen and Fuel Cell) Demonstration Project – Current Status Report

Jinichi Tomuro, Hydrogen Project Office, Engineering Advancement Association of Japan (ENAA), Tokyo, Japan

JHFC (Japan Hydrogen and Fuel Cell Demonstration) Project originally started in 2002 as a Japanese national project to demonstrate the effectiveness of fuel cell vehicles and hydrogen infrastructure for them. Its first phase lasted for 4 years and succeeded by its second phase which was expected to last for 5 years. Now the project entered into the last fiscal year. Under the JHFC project, 14 hydrogen refueling stations are operated and approximately 60 FCV's are on the public road. Up to the end of 2009, the cumulative driving distance reached 1.4 million kilometers and 29 tons of hydrogen had been refueled. In this paper, the author describes the results obtained from the project mainly in the field of hydrogen infrastructures. It includes various lessons learned, technical data and so on. Unfortunately the market introduction of FCV's seems to require a few more years. The Japanese government and the related industrial organizations began to consider the next demonstration project. Although the project scheme in the future is not clearly fixed yet, the author will try to introduce the upcoming project.

See more of this Session: Hydrogen Storage System Engineering and Applications: Infrastructure, Delivery Systems and Demonstration Systems
See more of this Group/Topical: Topical 8: Hydrogen Production and Storage

(26d) Development of Hydrogen Storage Technique for Fuel Cell Vehicle

Kenji Komiya, Fuel Cell System Development Div., Higashifuji Technical Center, Susono, Japan

Hydrogen is drawing attention as a clean and sustainable source of energy, and fuel cell vehicles that use hydrogen as fuel have the potential to become the ultimate automobiles of the future. However, since the energy density per volume of hydrogen fuel is extremely small (1/3000 of gasoline), it is important to develop technology to enable the storage of hydrogen in a more compact form. We will report development of hydrogen storage techniques for fuel cell vehicle including 70MPa CFRP high-pressure tank system and Multi-cylinder tank system.

See more of this Session: Hydrogen Storage System Engineering and Applications: Infrastructure, Delivery Systems and Demonstration Systems
See more of this Group/Topical: Topical 8: Hydrogen Production and Storage

Development of Hybrid Hydrogen Tank for Fuel Cell Vehicles

Shigeru Tsunokake[1], Tatsuya Fuura[1], Kazuya Takahashi[2], Yoshiki Sakaguchi[2], Etsuo Akiba[3], Kohta Asano[3], Hiroyuki Mitsuishi[4], Masanori Monde[5] and Yuichi Mitsutake[5]

(1)Japan Metals & Chemicals Co.,Ltd.(JMC), Tokyo, Japan, (2)SAMTEC CORP., Osaka, Japan, (3)National Institute of Advanced Industrial Science and Technology (AIST), Tsukuba Ibaraki, Japan, (5)Japan Automobile Research Institute (JARI), Ibaraki, Japan, (6)Saga University, Saga, Japan

Introduction

Hydrogen attracts attention as clean energy of the next generation. In order to put a hydrogen energy system to practical use, the technical development of safe and efficient storage/transportation media is on of the most important subjects. On board hydrogen tank which is under development can be classified into following two types; 1) high pressure hydrogen tank, 2) metal hydride storage tank. However, metal hydride storage tank do not meet the requirement for on board storage of hydrogen in weight densities. Many fuel cell vehicles adopt a high pressure hydrogen tank (35MPa, 70MPa), but have a problem to volumetric density now. This research is the development of a "hybrid hydrogen tank" which is a combination of type 3 high pressure hydrogen tank and metal hydride. "Hybrid hydrogen tank" is superior to a high pressure container of 70MPa in volumetric density. In this presentation, the development status of recent "hybrid hydrogen tank"[1)] [2)]will be introduced.

Target and Current state

This development projects were supported by New Energy and Industrial Technology Development Organization (NEDO). NEDO sets the target shown in Table 1 about the development goal in a hybrid tank.

Table 1. Target of Hybrid Hydrogen Tank System (Hydrogen: 5kg)

	2010(Current state)	2015(Target)	2020(Target)
Cost Target	¥5,000,000	¥2,000,000	¥200,000
Weight density Target	2-2.5wt%	3-4wt%	4-5wt%
Volume density Target	4.0kg/100L (Inner Volume)	5.0kg/100L (Inner Volume)	7kg/115L (Inner Volume)

Hybrid Tank

Hybrid tank is composed of type 3 high pressure hydrogen tank and MH cartridge.

The absorption of hydrogen by metal hydride is an exothermic reaction and desorption is an endothermic reaction, therefore it is necessary for the hybrid tank to have the function of the heat exchanger. MH cartridge consists of Metal Hydride and a heat exchanger. The structure of the hybrid tank is shown in figure 1 and the cut model of the hybrid tank is shown in figure 2.

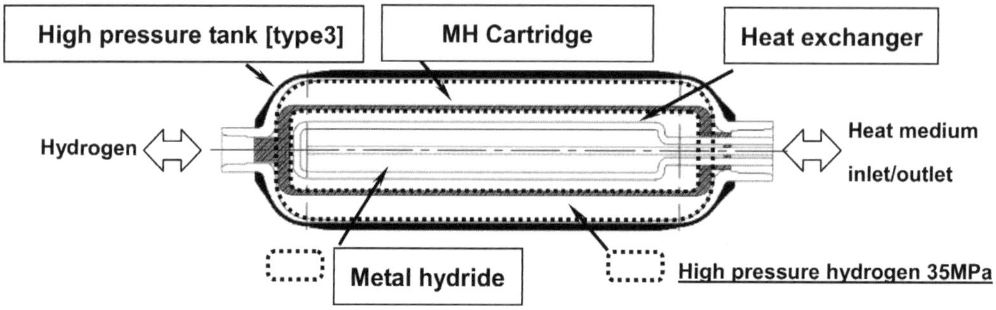

Figure 1. Hybrid tank structure

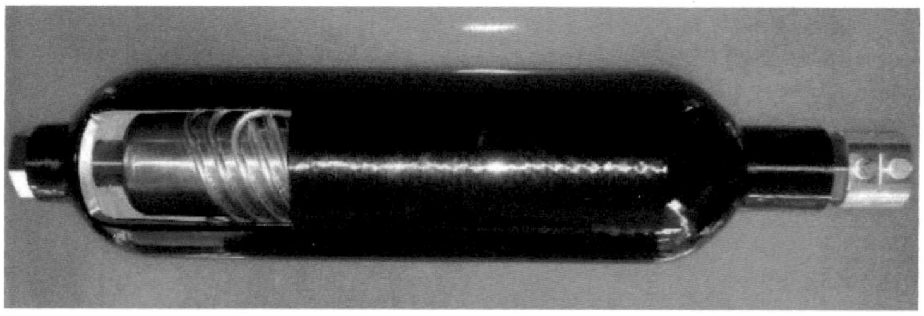

Figure 2. Cut model of hybrid tank

The specification of the hybrid tank is shown in table 2. The composition of metal hydride is V-5at%Ta-15at%Cr. The equilibrium pressure is 1.3MPa at 0deg.C. The value of the heat of formation for V-5at%Ta-15at%[3] is -30.0[kJ/mol H2].

Table 2. The specification of the hybrid tank

Inner volume	30.5 [L]
Total weight	61[kg]
(1) Type-3 tank weight	21[kg]
(2) heat exchanger weight	18.4[kg]
(3)Metal hydride weight	21.6[kg]
Hydrogen volume	1.1[kg]
Weight density	1.8[wt%]
Volume density	36[g/L]

Safety Test

We carried out a test of the safety along JARI-S001 which is standard of the safety of the type-3 container for Fuel cell vehicle. The safety test result is shown in table 3. Burst test, hydraulic cycle test , environmental cycle test and fire test was passed. But drop test was failure. The hybrid tank was heavier than the high pressure container (type-3), the damage was big. The strain for hybrid storage tanks was increased about 14%. For it the strain for the high pressure container (type-3) was increased about 6%. A fall of the performance of cycle test was predicted.

Table 3. Safety test result

Test	Test Result	Pass or Failure
Burst Test	Burst Pressure is 112MPa.	>78.75MPa Pass
Hydraulic Cycle Test	Leak was detected at 25,224 cycles.	>11,250 cycles Pass
Environmental Cycle Test	No Leakage	Pass
Fire Test	No Explosion (PRD activated)	Pass
Drop Test	-	Failure

Hydrogen charge test

We carried out hydrogen charge test of hybrid tank. The test carried out at Hydrogen & FCV Safety Evaluation Facility (Hy-SEF). The cooling medium temperature is -20deg.C and the flow rate is 60 L/min. The hydrogen pressure is 35 MPa. The test result is shown in figure 4.

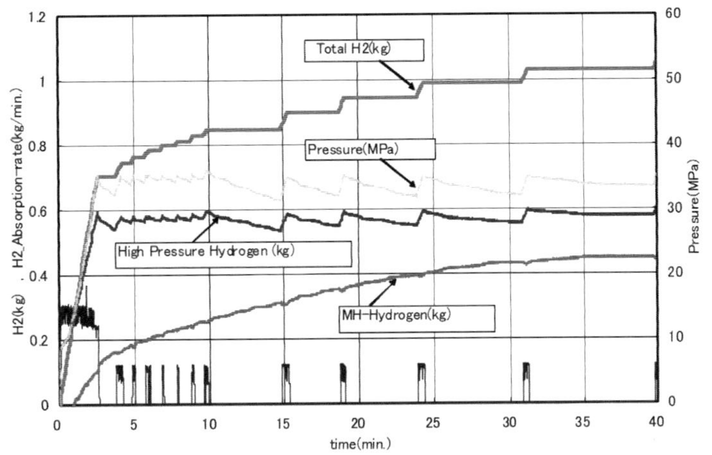

Figure 4. Hydrogen charge test

Acknowledgment

This development projects were supported by New Energy and Industrial Technology Development Organization (NEDO). This research has been conducted as a part of the "Development for Safe Utilization and Infrastructure of Hydrogen Research and Development of Vehicle Components" and "Development of Technologies for Hydrogen Production, Delivery and Storage Systems".

References

1 D.Mori et al. "High-pressure Metal Hydride Tank for Fuel Cell Vehicles",2005 MRS Spring Meeting

2 N.Takeichi et al./International Journal of Hydrogen Energy 28(2003) 1121-1129

3 NEDO "Development for Safe Utilization and Infrastructure of Hydrogen Research and Development of Vehicle Components" Report 2005

(26f) Testing of Hydrogen Storage Tank Based On 8 Kg NaAlH4

Oliver Metz, José M. Bellosta von Colbe, Gustavo A. Lozano, Julian Jepsen, Martin Dornheim and Thomas Klassen, Institute of Materials Research, Materials Technology, GKSS Research Centre Geesthacht, Geesthacht, Germany

Abstract

In the frame of the EU project STORHY (2004-2008) [1] an 8 kg-$NaAlH_4$ [2] based hydrogen storage tank has been designed and built. In parallel a new laboratory for storage tank testing and characterization has been constructed. According to the safety classification, this laboratory is divided in three parts: 1st part is a explosion proof chamber, 2nd part is a hydrogen certified laboratory and the 3rd part is used for the thermal auxiliary equipment. This paper will present basic features of the experimental testing facility, the storage tank investigated in this study as well as experimental results concerning the performance of the storage tank under different pressure/temperature and flow conditions like measured hydrogen charging and release rates as well as capacities measured under different conditions. An outlook on an improved second generation storage tank which is currently under development in the frame of the EU project NESSHY is given.

References:

[1] Ranong CN, Hohne M, Franzen J ,Hapke J, Fieg G , Dornheim M , Eigen N, von Colbe JMB, Metz O, accepted by CHEMICAL ENGINEERING & TECHNOLOGY, 32 (8): 1154-1163 AUG 2009 [2] Eigen, N; Keller, C; Dornheim, M; Klassen, T; Bormann, accepted by R, SCRIPTA MATERIALIA, 56 (10): 847-851 MAY 2007

See more of this Session: Hydrogen Storage System Engineering and Applications: Infrastructure, Delivery Systems and Demonstration Systems
See more of this Group/Topical: Topical 8: Hydrogen Production and Storage

(100a) Adsorbents for Hydrogen Delivery

Pascal Tessier[1], Frederic Barth[2], Sylvain Houle[3], Yudong Chen[1] and Richard Chahine[4], (1)Delaware Research and Technology Center, Air Liquide, Newark, DE, (2)Air Liquide Hydrogen Energy, Jouy-en-Josas, France, (3)Air Liquide Canada, Boucherville, QC, Canada, (4)Hydrogen Research Institute, Trois-Rivieres, QC, Canada

Hydrogen storage systems based on adsorbents are analyzed starting with a review of adsorbents and their properties. The storage capacity is dependent not only on the gravimetric capacity of adsorbents but also on their bulk density, the tank weight, and other parameters related to their operation. Since most adsorbents have a high capacity at low temperature, it is necessary to operate tanks at low temperature and preferable to fill them with cold hydrogen. Filling systems that can accomplish this are discussed. The potential performance of adsorbent-based storage and delivery systems are compared to commercial technologies.

See more of this Session: Hydrogen Storage System Engineering and Applications: Materials and Storage Risk Reduction
See more of this Group/Topical: Topical 8: Hydrogen Production and Storage

(100b) Overview: Development and Safety Test of Low Pressure and Hybrid Metal Hydride Tanks Under Japanese National Projects

Etsuo Akiba[1], Kohta Asano[2], Yoshinori Tsushio[3], Hiroyuki Mitsuishi[4], Kazuya Takahashi[5], Yoshiki Sakaguchi[5], Tatsuya Fuura[6] and Shigeru Tsunokake[7], (1)Etri, National Institute of Advanced Industrial Science and Technology (AIST), Tsukuba Ibaraki, Japan, (2)National Indtitute of Advanced Industrial Science and Technology (AIST), Tsukuba Ibaraki, Japan, (3)Mazda, Japan, (4)Japan Automobile Research Institute (JARI), Japan, (5)SAMTEC, Japan, (6)R&D Department, Japan Metals & Chemicals (JMC), Tokyo, Japan, (7)Japan Metals & Chemicals (JMC), Japan

Hydrogen storage materials contain hydrogen higher volume density than compressed gas and liquid hydrogen. Under Japanese national projects, there have been investigated hydride tanks with maximum pressure below 1 MPa (low pressure metal hydride tank) and those with much higher pressure such as15-35 MPa (high pressure metal hydride tank or hybrid tank). For the tank systems to be applied to real applications, not only performances of the system but also safety issues are critical. In this presentation, both performances and safety test of the tanks developed under the national projects will be introduced.

See more of this Session: Hydrogen Storage System Engineering and Applications: Materials and Storage Risk Reduction
See more of this Group/Topical: Topical 8: Hydrogen Production and Storage

(100c) Mitigation Technologies for Hydrogen Storage Systems Based On Reactive Solids

Daniel E. Dedrick[1], Michael P. Kanouff[1], Joseph G. Cordaro[1] and Yehia (John) F. Khalil[2], (1)Sandia National Laboratories, Livermore, CA, (2)Chemical Engineering Group, Physical Sciences Department, United Technologies Research Center (UTRC), East Hartford, CT

This paper describes mitigation technologies that are intended to enable the deployment of advanced hydrogen storage technologies for early market and automotive fuel cell applications.

Solid State hydrogen storage materials provide an opportunity for a dramatic increase in gravimetric and volumetric energy storage density. Systems and technologies based on the advanced materials have been developed and demonstrated within the laboratory [1,2], and in some cases, integrated with fuel cell systems. The R&D community will continue to develop these technologies for an ever increasing market of fuel cell technologies, including, forklift, light-cart, APU, and automotive systems.

Solid state hydrogen storage materials are designed and developed to readily release, and in some cases, react with diatomic hydrogen. This favorable behavior is often accomplished with morphology design (high surface area), catalytic additives (titanium for example), and high purity metals (such as aluminum, Lanthanum, or alkali metals). These favorable hydrogen reaction characteristics often have a related, yet less-desirable effect: sensitivity and reactivity during exposure to ambient contamination and out-of-design environmental conditions. Accident scenarios resulting in this less-favorable reaction behavior must also be managed by the system developer to enable technology deployment and market acceptance.

Two important accident scenarios are identified through hazards and risk analysis methods. The first involves a breach in plumbing or tank resulting from a collision. The possible consequence of this scenario is analyzed though experimentally based chemical kinetic and transport modeling of metal hydride beds. An advancing reaction front between the metal hydride and ambient air is observed to proceed throughout the bed. This exothermic reaction front can result in loss of structural integrity of the containing vessel and lead to un-favorable overheating events. The second important accident scenario considered is a pool fire or impinging fire resulting from a collision between a hydrocarbon or hydrogen fueled vehicle. The possible consequence of this scenario is analyzed with experimentally-based numerical simulation of a metal hydride system. During a fire scenario, the hydrogen storage material will rapidly decompose and release hydrogen at high pressure. Accident scenarios initiated by a vehicular collision leading a pipe break or catastrophic failure of the hydride vessel and by external pool fire with flame engulfing the storage vessel are developed using probabilistic modeling. The chronology of events occurring subsequent to each accident initiator is detailed in the probabilistic models.

Technology developed to manage these scenarios includes: (1) the use of polymer supports to reduce the extent and rate of reaction with air and water, (2) thermal radiation shielding. The polymer supported materials are demonstrated to provide mitigation of unwanted reaction while not impacting the hydrogen storage performance of the material. To mitigate the consequence of fire engulfment or impingement, thermal radiation shielding is considered to slow the rate of decomposition and delay the potential for loss-of-containment. In this paper we explore the use of these important mitigation technologies for a variety of accident scenarios.

[1]D. Dedrick, "Heat and mass transport in metal hydride based hydrogen storage systems", HT2009-88231 Proceedings of HT2009 ASME Summer Heat Transfer Conference July 19-23, 2009, San Francisco, CA [2]D. Mosher, "High Density Hydrogen Storage System Demonstration Using $NaAlH_4$ Complex Compound Hydrides, 2007 DOE H2 Annual Peer Review, Arlington, VA May 16, 2007

See more of this Session: Hydrogen Storage System Engineering and Applications: Materials and Storage Risk Reduction
See more of this Group/Topical: Topical 8: Hydrogen Production and Storage

(100d) Potential Diffusion-Based Failure Modes of Hydrogen Storage Vessels for On-Board Vehicular Use

Yehia F. Khalil, Chemical Engineering Group, Physical Sciences Department, United Technologies Research Center (UTRC), East Hartford, CT, Norman L. Newhouse, Lincoln Composites, Inc., Lincoln, NE, Kevin L. Simmons, Energy Materials Division, Pacific Northwest National Laboratory, Richland, WA and Daniel E. Dedrick, Sandia National Laboratories, Livermore, CA

Hydrogen-fueled fuel cell vehicles offer an environmentally attractive technology alternative to reduce greenhouse gas emissions driven, in part, by the transportation sector. The use of hydrogen/fuel cell integrated systems could also benefit other mobile applications such as forklifts and yard tractor utility vehicles. This research aims at quantifying the risks associated with gaseous diffusion through the walls of on-board vessels containing hydride materials as hydrogen storage media. The storage vessels of interest to this study include Type-III (made of carbon fiber reinforced composite wrapping with an inner metallic liner) and Type-IV (same as Type-III but with a polymeric liner instead of the metallic liner). In Type-III vessels, the liner could be aluminum or stainless steel and in Type-IV, high density polyethylene (HDPE) is commonly used as the liner material. Two failure modes related to gaseous diffusion are considered in this research. The first failure mode represents diffusion of the contained hydrogen gas through the vessel walls to the environment. Hydrogen could also leak through the vessel's welds, end fittings such as O-rings or other seals. The severity of consequences of this potential failure mode varies depending on the degree of confinement in which the permeated hydrogen will gradually accumulate, hydrogen leak rate, liner thickness, liner temperature, vessel internal pressure and the proximity to an ignition source. The second failure mode represents potential diffusion of oxygen from the surrounding air to the contained hydride inside the vessel. Most of the hydride materials, such as NaAlH4, MgH2, LiBH4, and LiAlH4, are air reactive (pyrophoric) and the severity of consequences depends on the amount of permeated oxygen inside the vessel as well as the pyrophoricity of the hydride. The consequential failures that may result from this failure mode may include over-temperature, vessel over-pressurization, or loss of containment integrity. It is also possible that the diffused oxygen gas may react with the contained hydrogen gas to form water vapor which would exacerbate the severity of hydride reactions. Also in Type-III vessels, there is a chance that the diffused oxygen could react with the metal liner and form oxides. In this work, consequence analysis using experiments and modeling is utilized to evaluate risk associated with these oxygen diffusion scenarios. Detailed event tree models are presented for each gaseous diffusion failure mode as an accident initiating event. The frequency of occurrence of the generated risk sequences will be quantified. The safety insights to be drawn from this study should assist risk analysts, OEM, Codes & Standards development teams, and other stakeholders in making risk-informed decisions with respect to identifying the optimum type of liner material, liner wall thickness and other design components of the hydride storage vessels for on-board vehicle applications.

(100e) Fundamental Reactivity Analysis of Hydrogen Storage Materials

David A. Tamburello[1], Charles W. James Jr.[2], Jose A. Cortes-Concepcion[2] and Donald L. Anton[2], (1) Computational Sciences, Savannah River National Laboratory, Aiken, SC, (2)Savannah River National Laboratory, Aiken, SC

Hydrogen is seen as the future automobile energy source due to its increased efficiency over conventional means and its ready utilization in fuel cell applications. In searching for ever higher gravimetric and volumetric density hydrogen storage systems, it is inevitable that higher energy density materials will be used. To make safe and commercially acceptable condensed phase hydrogen storage systems, it is important to understand quantitatively the risks involved in using and handling these materials to develop appropriate risk mitigation strategies to handle unforeseen accidental events. This will allow for the identification of potential risks and implementation of risk mitigation strategies.

A rigorous set of environmental reactivity tests have been developed based on modified testing procedures codified by the United Nations for the transportation of dangerous goods. Potential hydrogen storage materials—$2LiBH_4{\bullet}MgH_2$, NH_3BH_3, AlH_3, and $8LiH{\bullet}3Mg(NH_2)_2$—have been tested using these modified procedures to evaluate the relative risks of these materials coming in contact with the environment in hypothetical accident scenarios, including the identification of self-reactive substances, pyrophoric substances, and gas-emitting substances with water contact. An ignition event is most likely to occur if both a flammable concentration of hydrogen and sufficient thermal energy are available to ignite the hydrogen gas mixture. In order to predict hydride behavior for hypothesized accident scenarios, an idealized finite element model has been developed for a breached tank system resulting in dispersed hydride. Empirical thermodynamic calculations based on precise calorimetric experiments have been performed in order to quantify the energy and hydrogen release rates and to quantify the reaction products resulting from water and air exposure. Both thermal and compositional predictions have been made with respect to environmental scenarios identified as potential ignition events.

See more of this Session: Hydrogen Storage System Engineering and Applications: Materials and Storage Risk Reduction
See more of this Group/Topical: Topical 8: Hydrogen Production and Storage

(100f) Environmental Reactivity of Solid-State Hydrogen Storage Systems: Fundamental Testing and Evaluation

Charles W. James Jr., Jose A. Cortes-Concepcion, David A. Tamburello, Kyle S. Brinkman, Joshua R. Gray and Donald L. Anton, Savannah River National Laboratory, Aiken, SC

There is extensive research being conducted in developing new hydrogen storage systems that have higher gravimetric and volumetric capacities. However, less is understood about the hazards and risks associated with using and handling these materials. Therefore, it is critical to develop appropriate risk mitigation strategies to handle unforeseen events such as a breach-of-tank scenario in order to design commercially safe condensed phase hydrogen storage systems. A crucial aspect in risk identification is the development of rigorous environmental reactivity testing standards and procedures. One way to accomplish this is by adopting a modified testing approach from the United Nations[1] testing procedure for the transportation of dangerous goods to evaluate a potential hydrogen storage material reactivity to air and water. The modified U.N. procedures include identification of self-reactive, pyrophoric, and gas-emitting substances with water contact. The results of these tests for air and water contact sensitivity will be compared for such hydrogen storage candidates as $2LiBH_4{\cdot}MgH_2$, NH_3BH_3, $8LiH{\cdot}3Mg(NH_2)_2$, and AlH_3. The water contact tests are divided into two scenarios dependent on the hydride to water mole ratio and heat transport characteristics. Air contact tests were conducted to determine whether a substance will spontaneously react with air in a packed or dispersed form.

Additionally, thermodynamic calculations and substantiating calorimetric experiments were performed in order to quantify the energy released, energy release rates and to quantify the reaction products resulting from water and air exposure of various solid state systems. These calorimetric measurements, performed on the "small scale", were compared with standardized United Nations (UN) based tests for air and water reactivity and used to develop quantitative kinetic expressions for hydrolysis and air oxidation in these systems. Furthermore, insight gained from these U.N. tests as well as isothermal calorimetry will allow the formulation of additional tests geared toward safety engineering, numerical simulation, and integration.

[1] DOT/UN Doc., *Recommendations on the Transport of Dangerous Goods, Manual of Tests and Criteria*, 3rd Revised Ed., ISBN 92-1-139068-0, (1999).

See more of this Session: Hydrogen Storage System Engineering and Applications: Materials and Storage Risk Reduction
See more of this Group/Topical: Topical 8: Hydrogen Production and Storage

(151a) A System Modeling Approach for Light-Duty Vehicles and On-Board Hydrogen Storage Systems

Kristin L. Day, Matthew J. Thornton and Aaron D. Brooker, National Renewable Energy Laboratory, Golden, CO

Widespread adoption of hydrogen-fueled vehicles is heavily dependent on overcoming the challenges associated with on-board hydrogen storage. Identifying a means to store enough hydrogen capable of enabling a driving range greater than 300 miles within vehicle-related constraints is critical. High level vehicle, cost, and viability models integrated with detailed, low pressure, material-based hydrogen storage system models provide a means to assess the performance of a particular storage system technology in a vehicle context. The likelihood of this technology meeting the overarching vehicle system goals and thus becoming competitive with current vehicles can therefore be estimated. This paper discusses the methodology developed to evaluate the performance of various on-board hydrogen storage system technologies. Performance results focused on hydride, adsorbent, and chemical storage technologies will be presented and compared to vehicle-related packaging, cost, and performance requirements. Unique characteristics of each storage system – including gravimetric capacity, volumetric density, transient response time, full flow rate, and cost – will be identified and their effect on the viability of a theoretical vehicle will be assessed.

See more of this Session: Hydrogen Storage System Engineering and Applications: System Modeling
See more of this Group/Topical: Topical 8: Hydrogen Production and Storage

(151b) System Simulation Models for On-Board Hydrogen Storage Systems

Sudarshan Kumar[1], Mandhapati Raju[1] and V. Senthil Kumar[2], (1)Chemical Sciences and Material Systems Laboratory, General Motors Global R&D, Warren, MI, (2)India Science Laboratory, General Motors Global R&D, Bangalore, India

System simulation models for automotive on-board hydrogen storage systems provide a measure of the ability of an engineered system and storage media to meet system performance targets. Thoughtful engineering design for a particular storage media can help the system achieve desired performance goals. This paper presents system simulation models for two different advanced hydrogen storage technologies – a metal hydride system and a cryoadsorption system. Sodium alanate and super-activated carbon are employed as representative storage media for the metal hydride and the cryoadsorbent systems respectively. Lumped parameter models incorporating guidance from detailed transport models are employed in building the system simulation models.

Simulation results to test the storage systems' ability to meet fuel cell demand for different drive cycles and varying operating conditions are presented. Systems are engineered to provide the ability to refuel a vehicle in a short time guided by DOE targets. Gravimetric and volumetric hydrogen densities are computed for the engineered systems and compared to the DOE system goals. Finally, sensitivity of system performance to important design parameters is presented.

See more of this Session: Hydrogen Storage System Engineering and Applications: System Modeling
See more of this Group/Topical: Topical 8: Hydrogen Production and Storage

(151c) System Modeling Methodology and Analyses for Materials-Based Hydrogen Storage

Daniel Mosher, Jose Miguel Pasini and Bart van Hassel, United Technologies Research Center, East Hartford, CT

In the global efforts to develop advanced materials-based hydrogen storage, the various on-board reversible hydrides, adsorbents and chemical storage candidate materials and systems each have their individual strengths and weaknesses. An overarching challenge in associated research and development is to devise material/system architectures which satisfy all requirements for viability in a particular application area, such as light duty vehicular transportation. System modeling at the level which encompasses not only the storage material and vessel/reactor, but also integration with a fuel cell and balance-of-plant components, provides a more complete assessment of viability and guides options for improvement. The current work will cover the methodology developed for conducting such system modeling consistently across multiple organizations and will present performance results from studies focused on reversible hydride systems. Connecting this high level modeling to more detailed finite element design simulations will be one aspect of the framework approach discussed. The complex hydride $NaAlH_4$ is representative of novel materials under development and will be used as the basis for properties, such as temperature dependent kinetics, which influence the integrated system configurations and component sizing. While system charging is included, emphasis is placed on the discharge step in the storage process, interrogated through both simplified operation conditions (ex. constant fuel rate demand) and drive cycle transients. Comparisons of performance relative to requirements, including effective gravimetric capacity, effective volumetric density and energy utilization, will be given for the baseline material and for sensitivity studies on key material properties.

See more of this Session: Hydrogen Storage System Engineering and Applications: System Modeling
See more of this Group/Topical: Topical 8: Hydrogen Production and Storage

(151d) On-Board and off-Board Performance of Hydrogen Storage Options

Rajesh K. Ahluwalia[1], Thanh Q. Hua[2] and J.-K. Peng[2], (1)Nuclear Engineering Division, Argonne National Laboratory, Argonne, IL, (2)Nuclear Engineering, Argonne National Laboratory, Argonne, IL

A number of physical and materials-based hydrogen storage options are being developed to simultaneously meet the vehicular targets for gravimetric and volumetric capacity, cost, efficiency and greenhouse gas emissions, durability and operability, fuel purity, and environmental health and safety. The purpose of this paper is to evaluate the currently available physical, complex metal hydride, sorbent, and chemical hydrogen storage methods for their potential to meet these targets.

Our analyses show that hydrogen stored as a compressed gas (cH_2) at 350 bar in Type III or Type IV tanks cannot approach the near-term (2010) volumetric target of 28 g/L. Increasing the storage pressure to 700 bar improves the volumetric capacity (without reaching the target value) but degrades the gravimetric capacity. Dormancy and hydrogen loss are major concerns if hydrogen is stored as a cryogenic liquid (LH_2) in insulated low-pressure tanks, even if advanced vapor-cooled heat shield concepts are deployed. These problems can be mitigated by storing cryogenic hydrogen in pressure-bearing insulated tanks (the cryo-compressed option, CcH_2). Our analyses confirm that CcH_2 can also meet the near-term and intermediate-term (2015) targets for gravimetric and volumetric capacities. However, the cost of the cryo-compressed system is projected to be 3-4 times the target cost of $4/kWh, and the well-to-tank efficiency for the common pathway of producing hydrogen by central steam methane reforming (SMR) and subsequent hydrogen liquefaction is only about 41%, significantly lower than the efficiency target of 60%.

We have evaluated an on-board regenerable complex metal hydride ($NaAlH_4$, sodium alanate) system and an off-board regenerable metal hydride (AlH_3, alane) system. Our analyses show that, because of the large heat of reaction (ΔH = 37 kJ/mol-H_2) and slow hydrogenation and dehydrogenation kinetics, the doped sodium alanate system needs a combustor for releasing hydrogen, a bulky heat exchanger for refueling in 3-5 min, and a hydrogen buffer tank to handle startup and rapid demand transients. On the other hand, our analyses show that alane is an attractive hydrogen carrier (10 wt% H_2 content, ΔH = 6.7 kJ/mol-H_2), if it can be prepared and used as a slurry with >50% solids loading and a volume-exchange tank concept is developed that can store fresh and spent slurry in the same tank with a movable partition. Regenerating AlH_3 is a major problem, however, since it is metastable and it cannot be directly regenerated by reacting the spent Al with H_2. We have analyzed different indirect methods of regenerating AlH_3, including the organometallic route. We have constructed a flowsheet for this three-step method in which AlH_3 is formed as a stable adduct to an amine, the adduct is destabilized by transamination, and the AlH_3 is released by the catalytic decomposition of the transaminated adduct by decomposing under vacuum. We concluded that fuel cycle efficiencies in 40-55% range are possible with this organometallic route, if low-grade waste heat is available for the decomposition steps and the electricity requirements for the vacuum distillation steps match the estimates.

We have analyzed two sorbent-based hydrogen storage systems, one using AX-21, a high surface-area superactivated carbon, and the other using MOF-177, a metal organic framework material, Zn_4O(1,3,5-benzenetribenzoate) crystals. We found that moderate pressures and cryogenic temperatures are needed for reasonably high storage capacities, and that a temperature swing must be imposed in addition to the pressure swing to desorb hydrogen. Using off-board liquid N_2 for removing the heat of sorption is not very attractive because of the poor thermal conductivity of the sorbent bed and the unfavorable fuel cycle (well-to-tank) efficiency. We also evaluated the adiabatic refueling option, in which the sorbent bed is cooled evaporatively by charging it with liquid hydrogen, and warmed hydrogen is recirculated during discharge to supply the heat of desorption. The results from our analyses indicate higher storage capacities with MOF-177 than with AX-21, but these capacities are lower than those available with the cryo-compressed option.

We have also evaluated several exothermic and endothermic chemical methods of storing hydrogen. Releasing hydrogen by hydrolysis of sodium borohydride ($NaBH_4$, SBH) has been demonstrated in the laboratory and on-board an experimental vehicle. Our analysis of the SBH system pointed to a number of shortcomings, including the limited solubility of the product of reaction ($NaBO_2$) in the aqueous alkaline medium, difficulty in rejecting the large amount of heat that is liberated in the exothermic reaction ($\Delta H = 52.5$ kJ/mol-H_2), and the low well-to-tank efficiency in regenerating $NaBH_4$ from $NaBO_2$. We concluded that, for reasonable fuel cycle efficiencies, formation of B–O bonds should be avoided. We have evaluated the option of using an organic liquid carriers (LCH2), such as N-ethylcarbazole, which can be dehydrogenated thermolytically on-board a vehicle and rehydrogenated efficiently in a central plant by well-established methods and processes. Our analysis indicated low gravimetric (<3.5 wt% H_2) and volumetric (<25 g/L) capacities of this system, due to large ΔH (51 kJ/mol-H_2), pressure-limited equilibrium conversion, and slow dehydrogenation kinetics. As a result, the on-board system requires a mineral oil-based high-temperature coolant, a combustor, and a buffer H_2 storage tank for startup and rapid transients. Using reverse engineering, we have also determined the attributes that a prospective organic liquid carrier must have for such an on-board system to meet the H_2 storage targets.

Recently, there has been much interest in ammonia borane (NH_3BH_3, AB) as a hydrogen carrier because it has a high H_2 content (19.6 wt%), it undergoes exothermic thermolysis to release the H_2 ($\Delta H = 33$ kJ/mol-H_2), and it decomposes at relatively low temperatures. Like AlH_3, AB is thermodynamically unstable and its decomposition rate is independent of the backpressure. We are evaluating the concept of using AB dissolved in an organic liquid solvent (1-butyl-3-methylimadazolium, bmimCl) as an on-board storage medium. We have devised reactor configurations for controlling the peak temperature through heat transfer and recycle of the product stream. Like AlH_3, regenerating AB efficiently from the spent products is particularly challenging; it would be quite impractical if all three equivalents of H_2 are released from AB, allowing boron nitride (a ceramic) is allowed to form. We have constructed and analyzed flowsheets for off-board regeneration of spent AB using process chemistry developed by others. We have identified process steps that are energy intensive and are the largest contributors to inefficiencies if thiol or an alcohol is used to digest the spent AB. We concluded that, on a primary energy basis, regeneration efficiencies ranging from 17 to 35% are possible with these digestion schemes.

Figure 1 summarizes the volumetric and gravimetric capacities of the different on-board hydrogen storage options evaluated to date and compares them with the near-term, intermediate-term and ultimate targets. Figure 1 also compares the well-to-engine (WTE) efficiencies of these options for the pathway in which H_2 is produced in a central SMR plant. The paper will discuss others aspects of the storage options as well, including high-volume manufacturing cost projections, greenhouse gas emissions, durability and operability, and environmental health and safety issues (dormancy and hydrogen loss).

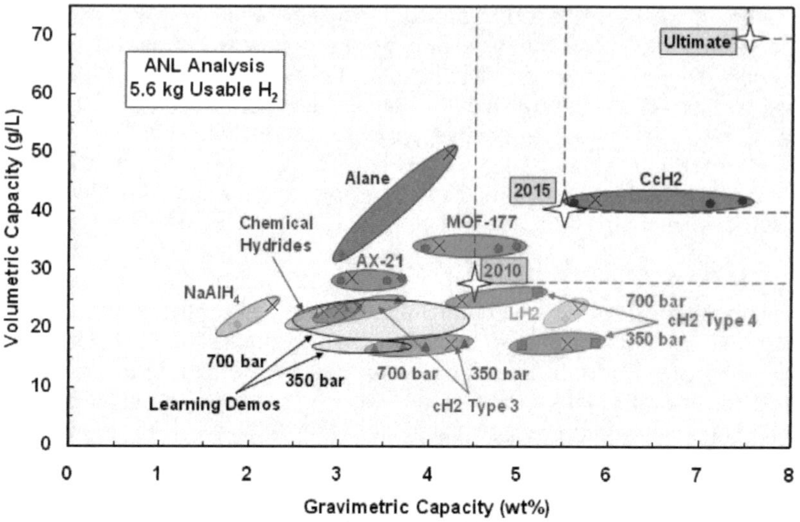

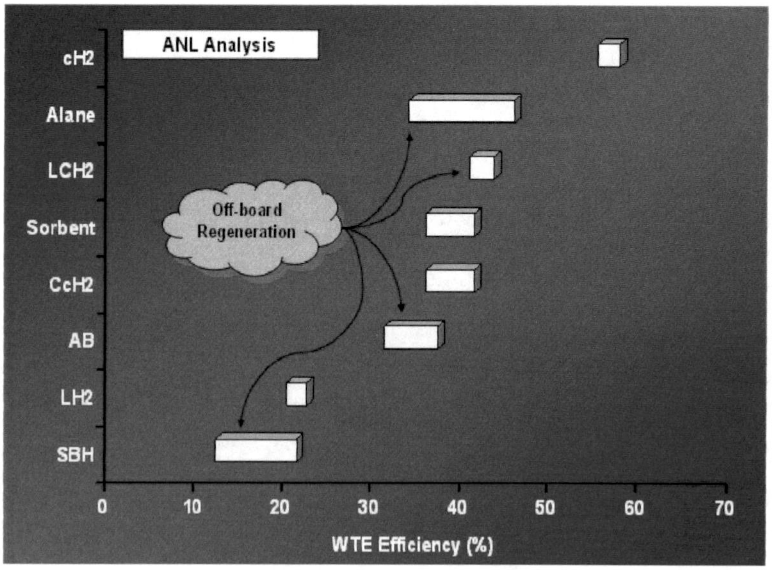

Figure 1 System capacities and fuel cycle efficiencies of different hydrogen storage options

See more of this Session: Hydrogen Storage System Engineering and Applications: System Modeling
See more of this Group/Topical: Topical 8: Hydrogen Production and Storage

(151e) Optimization of Metal Hydride Based Hydrogen Storage Bed Designs

Mandhapati Raju, General Motors Company, Warren, MI and Sudarshan Kumar, Chemical Sciences and Material Systems Laboratory, General Motors Company, Warren, MI

Design of metal hydride storage bed is a key factor that influences the storage capacity, gravimetric hydrogen storage density, and refueling time for automotive on-board hydrogen storage systems. The choice of storage bed design incorporating the heat exchanger, and corresponding design parameters is not obvious. A systematic study is presented to optimize the bed heat exchanger design using CFD modeling. Three different shell and tube heat exchanger designs are chosen. In the first design, metal hydride is present in the shell and heat transfer fluid flows through straight parallel cooling tubes placed inside the bed. The cooling tubes are interconnected by conducting fins. In the second design, the heat transfer fluid flows through helical tubes in the bed. The helical tube design permits use of a specific maximum distance between the metal hydride and coolant for removing heat during refueling. In the third design, the metal hydride is present in the tubes and the fluid flows through the shell. An automated tool is generated using COMSOL-MATLAB integration to arrive at the optimal geometric parameters for each design type. Using sodium alanate as the reference storage material, relative merits and demerits of each design type are analyzed in detail and a comparison of the gravimetric and volumetric hydrogen storage densities for the three designs is presented.

See more of this Session: Hydrogen Storage System Engineering and Applications: System Modeling
See more of this Group/Topical: Topical 8: Hydrogen Production and Storage

An NMR Study of Sodium Borohydride Hydrolysis in Concentrated Aqueous Solution

AIChE Meeting November 2010

Lin Yu[1], Amy M. Beaird[1], Ping Li[2], J. Karl Johnson[2], Rajasree Retnamma[3], Michael A. Matthews[1]

1. Department of Chemical Engineering, University of South Carolina

2. Department of Chemical and Petroleum Engineering, University of Pittsburgh

3. National Laboratory for Energy and Geology, Lisbon, Portugal

1.Introduction

Sodium borohydride is the most widely studied material among chemical hydrides and has been identified as a promising hydrogen storage media for portable applications, because of advantages such as high theoretical energy density (10.8wt%), low cost, safe handling and ease of H_2 release. Traditionally, studies focus on dilute sodium borohydride solutions with the aid of catalysts at room temperature. Various kinetic models for catalyzed hydrolysis have been reported accordingly. However, no kinetic data are available for non-catalyzed, concentrated $NaBH_4$ solutions. This work reports an experimental kinetic study of non-catalyzed sodium borohydride hydrolysis in highly concentrated solutions at elevated temperatures. The concentrations of boron containing species were characterized in-situ by a novel ^{11}B NMR technique. The effects of temperature and pH are reported.

2. Experimental

A Varian Inova 500 spectrometer operating at 160.487Hz for ^{11}B was used to acquire spectra in real time. The NMR operates in the temperature range of 25-90°C, with an accuracy of +/-0.1 °C. ^{11}B NMR spectra were acquired as proton-coupled, with 15s relaxation time and 8 scans. The NMR spectrometer was tuned to increase the signal-to-noise ratio each time before each measurement was carried out.

A concentric cylinder arrangement was used for obtaining real-time measurements of the rate of appearance of $NaB(OH)_4$. Sodium borohydride powder was placed in an 8 mm NMR tube, which was then sealed in a glove box. The 8mm tube was placed inside a 10mm NMR tube containing a standard solution of boric acid dissolved in D_2O. Prior to beginning the NMR scan, a predetermined volume of DI water was injected to the 8mm tube to dissolve the $NaBH_4$. Hydrogen bubbles were immediately seen after DI water was introduced. The concentric tubes were immediately placed in the spectrometer. The elapsed time between solution addition and first NMR spectrum acquisition was in the range of 180 to 240s.

The peak intensities of the reactant $NaBH_4$ and product $NaB(OH)_4$ were collected as a function of time. The concentrations of $NaBH_4$ and $NaB(OH)_4$ were quantified with reference to the standard boric acid solution, utilizing the fact that the concentration is proportional to the ratio of tube cross-sectional areas and the ratio of peak intensities. The data were obtained at temperatures of 25°C, 60°C and 80°C and at $NaBH_4$ concentrations of 10wt%, 15wt% and 20wt%. Additional data were obtained for the 20wt% $NaBH_4$ solution at 80°C, using NaOH at concentrations from initial 1wt% to 15wt% to adjust the pH. Another set of data were obtained for 10wt% $NaBH_4$ at 25°C, using acetic acid and HCl to adjust the pH in the range of 2.85-4.69.

3. Representative Results and Discussion

The effect of initial $NaBH_4$ concentration on the rate of formation of the hydrolysis is shown in Figure1.The data presented are concentration of the product $NaB(OH)_4$ (the formula equivalent to $NaBO_2 \cdot 2H_2O$ in excess water) at 25°C; It is shown that increasing the initial $NaBH_4$ concentration (10 to 20wt%) gives increasing product concentration. Comparison of the product concentration after 24 hours to that of 100% conversion is indicated by horizontal lines. As shown in Figure 1, the most concentrated solution results in lower conversion. In order to quantify the effects of initial $NaBH_4$ concentration on reaction rate at 25 °C, we took the derivative of the product concentration with respect to time. We find that the reaction rate drops quickly within first 5 hours; and it increases with the increasing initial $NaBH_4$ concentration from 10wt% to 20wt%.

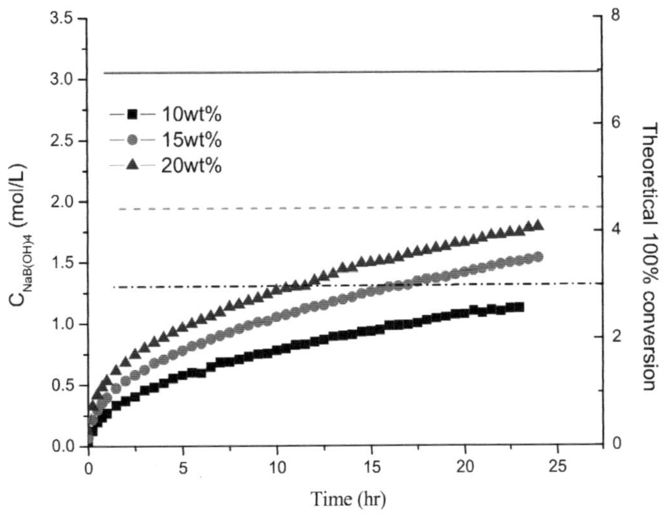

Figure 1. Effect of initial $NaBH_4$ concentration on formation of $NaB(OH)_4$ at 25°C

The effect of temperature within the range 25°C-80°C on the hydrolysis reaction has also been studied with an initial $NaBH_4$ concentration of 15wt%. Data will be presented during the presentation. It is found that at 80°C the reaction rate is the greatest; at around 10 hours, the $NaB(OH)_4$ has already reached theoretical value of 4.89mol/L for 100% conversion. However, at 25°C $NaB(OH)_4$ concentration only is 1.53 mol/L after 24 hours. Therefore, high temperature significantly improves the generation of product in the $NaBH_4$ hydrolysis. The reaction rate was observed to drop quickly within the first five hours and decrease gradually as the reaction continued. The initial reaction rate increases from 0.58 mol/L·hr to 6.73 mol/L·hr as temperature is increased from 25°C to 80°C.

To investigate the effect of increasing pH on the kinetics, alkaline solutions with different initial NaOH concentrations (1wt%, 5wt%, 10wt% and 15wt%) were prepared, with a constant initial $NaBH_4$ concentration of 20wt% at 80°C. NaOH inhibits the hydrolysis reaction. The inhibition effect is the least evident at 1wt% of NaOH, but this effect becomes more pronounced by increasing the concentration of NaOH from 5wt% to 15wt%. For the initial 15wt% NaOH, the hydrolysis rate is constant, very near zero.

The acceleration of the reaction with acid was accomplished by adjusting the initial pH. We used acetic acid and HCl to adjust the initial pH over the range from 2.85 to 4.69 to investigate the hydrolysis kinetics. It is found that with the same initial pH, acetic acid is a better accelerator than HCl at the same conditions.

To fit the experimental kinetic data, we fit the empirical power law model shown in equation (1). We assumed that both the $NaBH_4$ concentration and proton concentration will affect the hydrolysis rate:

$$R=\frac{dCNaB(OH)_4}{dt}=[NaBH_4]^{m}[H^+]^{n} \quad (1)$$

The kinetic parameters were obtained by a linear regression of the rate versus $NaBH_4$ concentration. The rate equation is based on the assumption of a constant water concentration during the hydrolysis reaction. We are able to correlate the measured pH with the product concentration under the same reaction conditions. For example, Table 1 shows the regression results from the left division algorithm in Matlab. The reaction order with respect to $NaBH_4$ decreases with increasing temperature from 25°C to 60°C; while the order with respect to proton concentration remains unaffected to temperature change in this range. The rate constant also shows an increase with rising temperature.

Table 1 Kinetic Parameters Determined by Linear Regression (neutral pH)

T	m	n	logk
25°C	2.8	1.1	7.5
60°C	1.9	1.1	9.5

We also performed a linear regression for the reaction order with respect to $NaBH_4$ and NaOH at 80°C, respectively. According to the result, the reaction rate model can be expressed by model equation (2):

$$R = k\,[NaBH_4]^{3.3}[NaOH]^{-0.92} \quad (2)$$

It is seen that the reaction rate increases with the $NaBH_4$ concentration and decreases with NaOH concentration. The negative dependence of reaction rate on the NaOH concentration is consistent with the inhibition effect of NaOH.

4. Conclusion

In this study, we investigated the non-catalyzed hydrolysis kinetics of concentrated $NaBH_4$ and NaOH stabilized solutions for hydrogen generation. The reactions are evaluated using ^{11}B NMR at 25, 60 and 80°C for 24 hrs. It is found that both high initial $NaBH_4$ concentration and high temperature are able to increase the reaction rate of $NaBH_4$ aqueous hydrolysis. Concentrated alkaline solutions will result in severe inhibition of the hydrolysis reaction. In contrast, acids accelerate the hydrolysis rate. We have also developed an empirical kinetic power-law model for the hydrolysis. It indicates that at higher temperature the dependence of hydrolysis reaction rate on $NaBH_4$ concentration decreases; the increasing

NaOH concentration decreases the reaction rate. Future work will take into account the water concentration change in modeling.

Ammonia Borane Dehydrogenation Always Generates Ammonia, How Much and How to Remove It?

Ahmad Al-Kukhun, Hyun Tae Hwang and Arvind Varma, School of Chemical Engineering, Purdue University, West Lafayette, IN

Hydrogen is a widely acknowledged clean fuel for PEM fuel cell (FC) vehicles and, Ammonia Borane (AB) is among the best hydrogen storage materials as it contains 19.6 wt.% and 150 kg/m3 H2 on weight and volume basis, respectively. Current methods to release hydrogen from AB include catalytic hydrolysis, thermolysis and hydrothermolysis [1]. As recently shown experimentally [2], all of these methods generate ammonia gas as a by-product of ammonia borane dehydrogenation. In addition to poisoning the anode catalyst site of PEM FC, ammonia can also cause ion exchange with protons in the polymer electrolyte. It has been reported that as low as 13 ppm NH3 can significantly decrease the FC performance, and that the degradation is irreversible for long-term exposure (15 hr) to 30 ppm NH3 [3]. Thus, to utilize any AB-based method to generate hydrogen for PEM FC applications, a technique for removing ammonia from the product gas stream is required.

In this study, for different operating conditions, we first quantified the ammonia generated from the more promising thermolysis and hydrothermolysis methods. Next, we evaluated different methods for ammonia removal, including membranes, absorption, adsorption, selective oxidation and cracking. The results show that, using a combination of these methods, ammonia can be removed completely. The optimized system configuration for ammonia removal was achieved to minimize the weight penalty of the added system.

[1] Diwan M., Hwang H. T., Al-Kukhun A., Varma A., "Hydrogen Generation from Noncatalytic Hydrothermolysis of Ammonia Borane for Vehicle Applications," AIChE Journal (in press, DOI 10.1002/aic.12240). 2010

[2] Hwang H. T, Al-Kukhun A., Varma A., "Hydrogen for Vehicle Applications from Hydrothermolysis of Ammonia Borane: Hydrogen Yield, Thermal Characteristics, and Ammonia Formation," Ind. Eng. Chem. Research (in press, DOI 10.1021/ie100520r). 2010

[3] Halseid, R., Vie, P. J. S., Tunold, R., "Effect of Ammonia on the Performance of Polymer Electrolyte Membrane Fuel Cells". J Power Sources 2006, 154: 343.

(218b) Critical Loading Level for Low-Temperature, Simultaneous Release of Two Equivalents of Hydrogen From Ammonia Borane Nanocomposites

Junshe Zhang[1], Yu Zhao[2], Daniel L. Akins[2] and Jae W. Lee[1], (1)Department of Chemical Engineering, The City College of New York, New York, NY, (2)Department of Chemistry, The City College of New York, New York, NY

Ammonia borane (NH3BH3, AB) has a gravimetric density of 196 g H2/kg AB; however, it slowly releases about 7 wt % hydrogen at temperatures below 85 degree Celsius. Although the dehydrogenation kinetic can be accelerated by compositing AB with mesoporous silica, the hydrogen density decreases significantly. The reduced density of nanocomposites can be offset by initiating the second hydrogen release step. However, this only occurs at low AB loading levels. Here, we report the critical loading level (or the maximum loading level) for the low-temperature, simultaneous release of two equivalents of hydrogen from MCM-41-AB nanocomposites. At a loading level of 0.15 (mass ratio of AB to matrix), 2.1 equivalents of hydrogen releases in 1 hour at 89.5 degree Celsius. Also, we find that nanoconfined AB (AB resides inside mesoporous channels) is amorphous and aggregated AB (AB resides outside mesoporous channels) has a tetragonal structure. Additionally, the first dehydrogenation of aggregated AB peaks at a low temperature compared with pristine AB. The critical AB loading level for nanocomposites involving MCM-41 and AB is found between 0.15 and 0.21. The findings improve understanding of mesoporous matrix-AB nanocomposites and are critical for design of novel AB nanocomposites.

See more of this Session: Fundamentals of Hydrogen Production: Reactions, Separations and Reactive Separations
See more of this Group/Topical: Environmental Division

(218c) Dihydrogen Catalysis Assisted by Molecular Hydrogen Reactions: Storage Linked Processes and Possible N2-Activation

Joseph Bozzelli, Chemistry and Environmental Science, New Jersey Institute of Technology, Newark, NJ and Rubik Asatryan, Chemistry and Env't Science, New Jersey Inst Tech, Newark, NJ

ABSTRACT. Density functional theory is employed in the context of molecular hydrogen assisted processes in order to disclose two types of reactions: (i) symmetric-syn¬chro¬nous double hydrogen-atom abstraction, and (ii) H2-mediated H-transfer. The reaction analysis is specifically applied to the heterogeneous decomposition of diazene, the most important intermediate of N2-activation processes.
KEYWORDS. Diazene, Dihydrogen Catalysis, Nitrogen Fixation, Nitrogenase, Hydrogen Storage.
While developing a comprehensive potential energy surface for hydrazine, N2H4 at CCSD(T)/CBS and CBS-QB3 levels of theory to uncover chemical activation reactions of amidogen radicals (NH2), we have revealed a stereoselective reaction pathway (Eq.1) directly associated to the gas-phase decomposition of cis-diazene, the most important substrate of nitrogenase enzyme and a key intermediate in heterogeneous catalytic processes involving elemental nitrogen and hydrogen, with fairly low gas-phase activation barrier of 22.5 kcal mol-1 (Fig.1): 1,2 H2 + cis-N2H2 ↔ N2 + H2 + H2 (1) Please Insert Fig.1. Here This reaction (Eq. 1) is the simplest possible model for the symmetric-synchronous double hydrogen-atom abstraction reaction from an H-donor as cis-diazene molecule. or: D2 + cis-N2H2 • 2 DH + N2 D–notation we employed here for clarity, and note that such an isotopic scheme explains the observed formation of HD during N2-reduction in the presence of D2 (vide infra). 3 Intriguingly, the reverse N2-activation reaction (Eq.-1) faces a barrier of only 76 kcal mol-1 which is almost 50 kcal mol-1 lower than the analogous bimo¬le¬cu¬lar nitrogen fixation reaction (ca.125 kcal mol-1).4 The reverse reaction is termolecular and is consequently, significantly un¬favored entropically. The reaction becomes viable at very high pres¬sures5 but would be feasible if the molecular hydrogen is partially bonded to a hetero-surface and the H-H bond is both polarized and weakened. Natural Bond Orbital (NBO) analysis of the reaction (1) at B3LYP/6-311++G(2d,2p) level of theory shows that N-H bonds in isolated cis-N2H2 are polarized (natural charges on H and N atoms are QH= +0.14e and QN= -0.14e, respectively). However, the electron density in the transition state ring (Fig.1) is almost completely delocalized, with partial polarization of newly formed H-H bonds (D-H in Eq.1): QH= +0.08e and QN= -0.07e. Terminal hydrogen atoms (D atoms) posses small negative charges, QD = -0.01e. Natural electron configuration of nitrogen atoms in TS is 2s0.782p1.77. As expected, 2p-orbital electrons are responsible for the partial electron donation from the isolated diazene (2s0.762p1.86) to the approaching hydrogen. Hence, promoting role of the initial activation of attacking H2 can be anticipated. In this report, we have studied analogous metalorganic systems and discuss several possible implications. The fixation of N2, generally considered as its reduction to NH3, is an important topic throughout the natural sciences studied over several decades; yet the detailed molecular mechanism remains an open issue.6-11 The N≡N bond is exceptionally strong (ca. 225 kcal mol-1) and hence the

first step activating this triple bond is extremely unfavorable despite the overall process of ammonia synthesis being exothermic by ca 22 kcal mol-1. Recent theoretical calculations at G2M(MP2)//MP2/6-31G** level suggest[4a] that the lowest barrier associating H_2 and N_2 into iso-diazene is 125.4 kcal mol-1, consonant with the MRCI(+Q)/aug-cc-pVQZ//FVCAS/aug-cc-pVQZ value of 5.39 eV.[4b] Chain reactions are therefore considered as the dominant gas-phase and surface mediated mechanisms for hydrogenation of nitrogen.[4] It is known, that very high pressures and temperatures (200 atm and 500oC) with iron-based catalysts are employed to hydrogenate nitrogen in the industrial Haber-Bosch process, which is believed to proceed via initial dissociative adsorption of reagents in a (Fe, or Ru based) catalyst surface. [4b] In contrast, the activation of N_2 occurs at ambient conditions in biological media, supported by enzymes, such as nitrogenase.[6] The con¬ven¬¬tional bio-catalytic mechanism considers binding of N_2 to the Fe-[H_2] comp¬lex, reduction of N_2 to diazene (diimide, N_2H_2) and hydrazine (N_2H_4) and release of NH_3.[5] Hydrazine as an inter¬me¬diate has been detected ex¬pe¬rimentally, while diazene, which is also proven to be a substrate for nitrogenase, [7] is not observed due to its high reactivity. [6b] It is further important to note that molecular hydrogen is always produced when nitrogenase reduces N_2 to NH_3 (a turnover cycle); evolution of hydrogen by nitrogenase is uniformly observed, even under high nitrogen pressures (e.g., 50 atm).[8] Most N_2 fixing bacteria contain hydrogenase, the another relevant enzyme, to recycle excessive hydrogen.[9] The equation for conversion of N_2 on nitrogenase is considered as $N_2 + (6+2x)H^+ + (6+2x)e^- \bullet 2NH_3 + xH_2$ (2) The value of x varies from 1.0 - 7.5 depending on the type of enzyme used; for main MoFe enzyme it is considered unity. [6,8,10] The nitrogenase also reduces other substrates such as C_2H_2, HCN and N_2O to C_2H_4, $CH_4 + NH_3$ and $N_2 + H_2O$, respectively. Note that hydrogen is not evolved during enzymatic reduction of acetylene and interestingly, C_2H_2 is specifically converted to cis-$C_2H_2D_2$ in presence of D_2O.[6] Iron plays a vital role in variety of enzymatic and heterogeneous catalytic processes.[6,9,11] We have therefore, studied several simple heterogeneous models based on iron-organic complexes to further evaluate the relevance of assisted by molecular hydrogen mechanisms. Calculations were performed at the B3LYP hybrid DFT level[12] using the Dunning-Hay double zeta basis set[13] and Los Alamos effective core potential[14] for Fe as implemented in Gaussian-03. [15] A Dresden/Stuttgart pseudo¬potential is also used to verify key results, not presented here. Different spin-less (S=0) and high-spin (S=4) neutral and ionic systems were considered for comparison. In this short report the reaction of hydrogen molecule with a straightforward $(CH_3)_2FeH_2$ model as a source of activated hydrogen atoms will be addressed. A five (trigonal bipyramidal) and a four (tetrahedral) coordinated Fe-complexes are also adopted, to generally model some features of the organometallic catalytic active sites. Structures similar to the gas-phase transition state (Fig.1) are observed for reactions between the organome¬tallic¬ model $(CH_3)_2FeH_2$ with diazene and CH_2NH intermediates (Fig2a,b), where a hydride of a tran¬si¬tion metal serves as a donor of the dihydrogen, a "H_2-catalyst": R_2FeH_2 + cis-N_2H_2 = $R_2Fe + H_2 + H_2 + N_2$ (3) $R_2FeH_2 + CH_2NH = R_2Fe + H_2 + H_2 + HCN$ (4) In Fe-organic models the hydridic H-atoms are pre-activated, and this substantially reduces the activation energy. The barrier height for the catalytic decomposition reaction of cis-N_2H_2 (Eq.3) with R≡CH_3 is fairly low (6.5 kcal mol-1) compared to the 22.5 kcal mol-1 barrier in the gas-phase. Thus, the ring-mechanism is expected to be functioning in both gas and condensed phase systems. Please Insert Fig.2. Here This mechanism dealing with different nitrogenase substrates and intermediates (such as diazene, HCN, N_2O, etc) may serve as a simple molecular model involving requisite evolution of hydrogen (an attribute to the turnover cycle of nitrogenase enzyme) and could explain why the activation process is inhibited by H_2.[8] We demonstrate that a similar ring- transition state endures in different environments, in a variety of neutral and ionic systems. This dihydrogen catalysis phenomenon can be possibly applied to the decomposition processes of high hydrogen content compounds. It opens also a variety of new pathways re¬le¬vant to a number of processes, including remote transfer of H-atoms supported by molecular hydrogen. 16,18 We report one further, me¬¬cha¬nistically important mechanism, viz., a new H2-mediated H-transfer reactions, which we believe may occur virtually in any systems. We demonstrated this mechanism on a simple Fe-S-Fe backbone organometallic model with neutral CO ligands (Fig.3), which is a rather common moiety in biological systems.16 Fig.3a demonstrates a regenerative cycle for hydrogen transfer from an S-H group to the coordinated N2H-group, the possible intermediate of N2-hydro¬ge¬na¬tion. Here, the reactant H2 molecule is a transferor of an H atom from a bridged S to the distal N atom. It can be considered a catalyst as it is chemically unchanged at the end of reaction, when only one hydrogen isotope is involved. A similar process holds in the case of H-transfer from a Fe-H group in unsaturated tetrahedral Fe1 center to the axially located N2H group in the trigonal bipyramidal structure (Fig3b), while formed (and yet adsorbed) N2H2–group transfers back to the unsaturated Fe1 center. Note, that hydrogenation (protonation with further electron attachment) of a bridged sulfur atom in nitrogenase models is shown to facilitate activation of N2 adsorbed on the proximate Fe-centers.10,17

See more of this Session: Fundamentals of Hydrogen Production: Reactions, Separations and Reactive Separations
See more of this Group/Topical: Environmental Division

(218d) Energetically Self-Sufficient Coproduction of Hydrogen and Formic Acid

Jorge Pena Lopez and Vasilios I. Manousiouthakis, Chemical Engineering Department, University of California, Los Angeles, Los Angeles, CA

The use of hydrogen has been proposed as a solution to the air quality deterioration in metropolitan areas and as an effective energy carrier for small and medium vehicles. This option has strengthened with the advancement in the development of fuel cells manufacturing and hydrogen storage. To technologically advance this alternative, new hydrogen production designs are to be developed. Several options exist for hydrogen production; one of the most explored has been the reforming of natural gas given that methane contains potentially two hydrogen gas molecules. The main problem with conventional natural gas reformers is the emission of carbon dioxide, a greenhouse whose emissions have led to an increase in its concentration in the atmosphere. Transforming carbon dioxide into a useful product prevents its emission into the atmosphere; however, an inadequate process design might largely increase the cost of producing hydrogen, making it uncompetitive with gasoline. In this work we present a comparison between three different hydrogen and formic acid coproduction process. The first design provides a baseline for the production of hydrogen with recovery of carbon dioxide downstream of the steam reforming process, where the carbon recovered in this way is utilized to produce formic acid. The second design recovers the carbon dioxide released at the source of energy generation, thus producing a larger amount of formic acid and eliminating carbon emissions to zero. A third design implements the commercially established formic acid production process into a design including hydrogen production. Given the structure and feedstock of this formic acid plant, the hydrogen production can no longer by carried out by a steam reforming process but rather from a combination of dry reforming and water gas shift. These three processes go through heat and power integration techniques that allow the maximum utilization of energy in favor of hydrogen production, maintaining a self-powered plant and maintaining the carbon dioxide emissions at zero. We show that there exists an optimal themodynamical and operational point for the operation of these plants.

See more of this Session: Fundamentals of Hydrogen Production: Reactions, Separations and Reactive Separations
See more of this Group/Topical: Environmental Division

(221a) Research and Development of Hydrogen Storage Materials for Lightweight Tanks

Yoshitsugu Kojima and Takayuki Ichikawa, Institute for Advanced Materials Research, Hiroshima University, Higashi-Hiroshima, Japan

Fuel cell vehicles (FCV) feature 35-70 MPa hydrogen storage tanks and can travel 620 km (pressure: 35 MPa, inner volume:171 L) -780km (pressure: 70 MPa, inner volume:156 L) on full tanks. The tank size of the vehicle is large compared to those of gasoline vehicles. Hydrogen can be stored in many different forms as compressed or liquefied hydrogen in tanks, as hydrogen adsorbed carbon materials, as hydrogen-absorbing alloys, as metal hydrides with light elements such as NaBH4, NaAlH4, MgH2, LiBH4 and metal nitrides, or as organic hydrides (methylcyclohexane, decalin)[1]. In this study, we will review our experimental results on metal hydride with high dissociation pressure (Ti1.1CrMn), light weight Li-Mg-N-H system, NH3-based materials and NaH-NH3BH3 composites. We found that the effective hydrogen capacity of Ti1.1CrMn exhibited the maximum value of 1.8 mass% in the pressure range of 33 MPa and 0.1 MPa at 296 K (dissociation pressure: 11 MPa) [2]. At the low temperature of 233 K, the hydrogen desorption capacity at 0.1 MPa was 1.6 mass%. The standard enthalpy difference of Ti1.1CrMn was -22 kJ/molH2. The absolute value was about 10 kJ/molH2 smaller than those of LaNi5 and Ti-Cr-V. The mixtures of LiH and Mg(NH2)2 with molar ratios of 8 to 3 were milled for 2-20 h under 1 MPa H2 pressure to synthesize the Li-Mg-N-H system by a planetary ball milling [3]. It was found that milling time had a remarkable effect on improving kinetics and then an optimum milling time was 20 h. The H2 desorption capacity close to 6 mass% was obtained after keeping for 8 h at 423 K. For the dehydrided Li-Mg-N-H system (vacuum, 473K, 24h), ball milling treatment was performed at 423 K for 5 h under 15 MPa of H2 pressure, leading to the H2 absorption capacity up to 5.8 mass% . Alkali metal hydrides (LiH, NaH and KH) reacted with NH3 at room temperature by the exothermic reactions [4, 5]. In the hydrogen generating reactions, the activation treatment by ball milling drastically improved the kinetics for the Li, Na, K systems. After these H2 generations, alkali metal amides were able to store H2 at 373-573 K, forming NH3 and the corresponding alkali metal hydrides under 0.5 MPa H2 flow for 4 h. H2 of 8.1 [H2/LiH+NH3] , 4.9 [H2/NaH+NH3] and 3.5 mass% [H2/KH+NH3] can be reversibly stored by this reaction. Sodium hydride-ammonia borane composite systems were prepared by ball milling. The excess hydride (NaH/NH3BH3= 2-3 /mol/mol) suppressed the foaming and ammonia gas release. The dehydrogenation peak temperature decreased with increase in the NaH content. For the composite materials of 3NaH/ NH3BH3, it reaches the dehydrogenation peak at 345 K, which is much lower than the dehydrogenation temperature of Na NH2BH3 (357K). When we filled the tank (inner volume: 100 L) with the Ti1.1CrMn alloy of 50 kg, the gravimetric and the volumetric hydrogen densities at 35 MPa and 298 K were calculated as 6.0 mass% and 3.0 kg/100 L, respectively. The volumetric hydrogen density provides 30 % more capacity than compressed hydrogen at 35 MPa (volumetric hydrogen density: 2.3 kgH2/100L) and the gravimetric and the volumetric hydrogen densities are similar to those of the Li-Mg-N-H system (packing ratio: 50%, packing density : 0.6 g/cm3, gravimetric hydrogen density: 5.8 mass%, volumetric hydrogen density: 3.5 kgH2/100L). We have calculated the gravimetric and the volumetric H2 densities of the LiH-NH3 system. At 70 MPa, the volume of a tank (100L) for gaseous hydrogen allows storage density of 3.9 kgH2. We assumed that the density of LiH was 0.78 g/cm3. When we filled the tank (internal volume 62 L) with liquid NH3 (density: 0.61 g/cm3) of 38 kg and the tank (internal volume 46 L, packing ratio 50 %) with LiH of 18 kg (LiH/NH3=1mol/1mol), the gravimetric and the volumetric H2 densities are calculated as 8.1 mass% and 4.2 kg/100L, respectively. The volumetric H2 density provides 1.8 and 1.1 times of compressed hydrogen at 35 and 70 MPa (3.9 kgH2), respectively. References [1]"Hydrogen Storage". In: J. Garche, C. Dyer, P. Moseley, Z. Ogumi, D. Rand, B. Scrosati, editors. Encyclopedia of Electrochemical Power Sources, Vol 3. Amsterdam: Elsevier; 2009. [2]Y. Kojima, Y. Kawai, S. Towata, T. Matsunaga, T. Shinozawa and M. Kimbara, J. Alloys Compd. 419 256 (2006). [3]Y. Wang, T. Ichikawa, S. Isobe, M. Tsubota, S. Hino, T. Nakagawa and Y. Kojima, Abstracts of MH2008, International Symposium on Metal Hydrogen Systems, June 24-28, Reykjavik, Iceland. [4]Y. Kojima, K. Tange, S. Hino, S. Isobe, M. Tsubota, K. Nakamura, M. Nakatake, H. Miyaoka, H. Yamamoto and T. Ichikawa, J. Mater. Res., 24, 2185 (2009). [5]H. Yamamoto, H. Miyaoka, S. Hino, H. Nakanishi, T. Ichikawa and Y. Kojima, Int. J. Hydrogen Energy, 34, 9760 (2009).

See more of this Session: Hydrogen Storage System Engineering and Applications: Applied Materials Development I
See more of this Group/Topical: Topical 8: Hydrogen Production and Storage

(221b) High Capacity Metal Hydrides – Development towards Scaled up Tanks and Economical Evaluation

José M. Bellosta von Colbe[1], Julian Jepsen[2], Gustavo Lozano[3], Oliver Metz[3], Thomas Klassen[3] and Martin Dornheim[3], (1)Institute of Materials Research, Materials Technology, GKSS Research Centre Geesthacht, Geesthacht, Germany, (2)Institute of Materials Research, Materials Technology, GKSS-Forschungszentrum Geesthacht GmbH, Geesthacht, Germany, (3)GKSS Research Centre Geesthacht, Geesthacht, Germany

Hydrogen storage seems to be a major issue on the way to hydrogen mobility and usage of hydrogen as an energy vector. High capacity metal hydrides and composites, like sodium alanate and the Reactive Hydride Composites (see talks from Lozano and Jepsen) offer promising solutions, especially for applications involving the use of high or medium temperature heat carriers. These materials, some of which are extremely new, need to be developed from the milligram scale in which they were first synthesized to the kg and even Ton scale in order to be made available as applied storage systems. This requires the investigation of properties that are not the first priority of researchers looking for new materials: heat capacity and conductivity, porosity, cyclability, and especially the issues involving aging and deactivation. In this talk, the activities and techniques used to investigate hydrides at GKSS will be described on the example of sodium alanate, but also including other hydrides. Moreover, the final criterion for an energy storage system is its economic viability. Results will be provided showing that hydrides are a cost – effective solution to the issue of hydrogen storage in the medium scale (tens of kg.)
Hydrogen storage seems to be a major issue on the way to hydrogen mobility and usage of hydrogen as an energy vector. High capacity metal hydrides and composites, like sodium alanate and the Reactive Hydride Composites (see talks from Lozano and Jepsen) offer promising solutions, especially for applications involving the use of high or medium temperature heat carriers. These materials, some of which are extremely new, need to be developed from the milligram scale in which they were first synthesized to the kg and even Ton scale in order to be made available as applied storage systems. This requires the investigation of properties that are not the first priority of researchers looking for new materials: heat capacity and conductivity, porosity, cyclability, and especially the issues involving aging and deactivation. In this talk, the activities and techniques used to investigate hydrides at GKSS will be described on the example of sodium alanate, but also including other hydrides. Moreover, the final criterion for an energy storage system is its economic viability. Results will be provided showing that hydrides are a cost – effective solution to the issue of hydrogen storage in the medium scale (tens of kg.) [1]

References [1] Jepsen, J. et al. submitted to Int. J. of Hyd. Ener.

See more of this Session: Hydrogen Storage System Engineering and Applications: Applied Materials

(221c) Sorption Kinetics and Catalysis of 1:1 $LiNH_2$:MgH_2

Donald Anton[1], **Christine J. Erdy**[1] and Joshua Gray[2], (1)Energy Security, Savannah River National Laboratory, Aiken, SC, (2)Materials Comp & Weld Tech, Savannah River National Laboratory, Aiken, SC

High-performance on-board hydrogen storage systems are increasingly recognized as critical to implementation of hydrogen fuel cells as clean, efficient automotive power plants [1]. Among the various storage system types: adsorbent materials, chemical hydrides and metal hydrides, the use of complex hydrides is one of the candidates because of their large gravimetric and volumetric storage capacities [2], on-board reversibility, and indefinite ambient temperature storage duration. The destabilized complex hydride system composed of 1:1 $LiNH_2$:MgH_2 is one of the leading candidates of hydrogen storage compositions. In 2007, Lu et al. showed that the reaction between MgH_2 and $LiNH_2$ in a 1:1 molar ratio resulted in a new nitride-binary light metal, LiMgN, for hydrogen storage applications [3]. The dehydrogenation reaction pathway is as follows: $LiNH_2+MgH_2 \rightarrow LiMgN+2H_2$ This 1:1 molar ratio of MgH_2 to $LiNH_2$ mixture has a theoretical hydrogen weight capacity of 8.2 wt%. The rehydrogenation and subsequent dehydrogenation process of LiMgN produces LiH, $Mg(NH_2)_2$, and MgH_2, as seen here: $LiMgN+2H_2 \leftrightarrow LiH+0.5Mg(NH_2)_2+0.5MgH_2$. In a study of more than 300 destabilization reactions using first principle density function theory (DFT), Alapati et al. predicted that the reaction of between MgH_2 and $LiNH_2$ in a 1:1 molar ratio was energetically favorable with an enthalpy of 31.9 kJ/molH_2 [4]. This enthalpy suggests that favorable sorption thermodynamics should be obtainable at temperatures in the range of 160°C to 260°C. Preliminary experiments reported in literature indicate that sorption kinetics are substantially lower than expected in this temperature range despite favorable thermodynamics. Systematic isothermal and isobaric sorption experiments were performed on these materials using a Sievert's apparatus to form a baseline data set by which to compare kinetic results. Various catalytic modifications were performed in attempts to increase the kinetics while lowering the sorption temperatures. In this paper, we investigate the base line sorption kinetics of the 1:1 $LiNH_2$:MgH_2 composition and the effects of catalyst on these rates, the temperature of initial hydrogen release and the amount of ammonia released from the catalyzed and catalyst-free materials. Possible implications for use of this material for hydrogen storage applications are summarized.

1. Pinkerton, F.E., Decomposition kinetics of lithium amide for hydrogen storage materials. Journal of Alloys and Compounds, 2005. 400(1-2): p. 76-82.

2. Matsumoto, M., et al., Hydrogen desorption reactions of Li-N-H hydrogen storage system: Estimation of activation free energy. Journal of Alloys and Compounds, 2007. 439(1-2): p. 358-362.

3. Lu, J., et al., Potential of Binary Lithium Magnesium Nitride for Hydrogen Storage Applications. The Journal of Physical Chemistry C, 2007. 111(32): p. 12129-12134.

4. Alapati, S.V., J.K. Johnson, and D.S. Sholl, Identification of Destabilized Metal Hydrides for Hydrogen Storage Using First Principles Calculations. The Journal of Physical Chemistry B, 2006. 110 (17): p. 8769-8776.

See more of this Session: Hydrogen Storage System Engineering and Applications: Applied Materials Development I
See more of this Group/Topical: Topical 8: Hydrogen Production and Storage

(221d) Engineering Advances in NaAlH4-Based Hydrogen Storage System

Bart A. van Hassel[1], Daniel A. Mosher[2], Jose Miguel Pasini[3], Mikhail B. Gorbounov[4], John E. Holowczak[5], Xia Tang[5] and Ronald J. Brown[5], (1)Physical Sciences, United Technologies Research Center, East Hartford, CT, (2)Applied Mechanics, United Technologies Research Center, East Hartford, CT, (3)System Dynamics & Optimization, United Technologies Research Center, East Hartford, CT, (4) Thermal Management, United Technologies Research Center, East Hartford, CT, (5)Ceramics, United Technologies Research Center, East Hartford, CT

The significant engineering challenges associated with developing lower-pressure, materials-based, hydrogen storage systems for hydrogen fuel cell light-duty vehicles are being addressed by focusing on the role that powder consolidation can play. NaAlH4 with 4 wt. % TiCl3 was selected as the model material. We focused on the changes in the physical (density and thermal conductivity) and mechanical properties (biaxial flexure strength) and on how this impacted the volumetric capacity of the hydrogen storage system. Both the thermal conductivity and the density of the ball milled material improved with applied pressure in an uniaxial press over the range of 14 MPa to 281 MPa. The thermal conductivity reached a value of 1.637 W/m/K, which was a factor seven higher than that of the unconsolidated powder. The volume of the material was reduced with 58% at the highest applied pressure. A method was developed for determining the strength of NaAlH4 pellets before and after hydrogen absorption and desorption cycles. It is based on a biaxial flexure test that was originally designed for determining the strength of green ceramic materials. The tests showed that the pellets were strong with a biaxial flexure strength of 1.4 kpsi which was unaltered over three studied hydrogen absorption/desorption cycles. The increased materials density did not affect the hydrogen absorption and desorption kinetics, which is important in order to benefit from the improved volumetric capacity. The new material properties of the compacted NaAlH4 were used in finite element modeling of a hydrogen storage system that targeted a fast refueling time. The results clearly show an improvement of the volumetric capacity of the system by powder consolidation but the gravimetric capacity remains below target, as expected. A system level study of a light-duty vehicle with such a hydrogen storage system is required in order to determine whether the amount of hydrogen stored in the pore volume of the sodium alanate will still be enough to enable one cold start from room temperature to its operating temperature (120-140°C) or that a buffer volume needs to be installed. While it is recognized that a sodium alanate based hydrogen storage system has its limitations, it has been demonstrated that powder consolidation can address some of those limitations by improving the thermal conductivity and volumetric capacity.

See more of this Session: Hydrogen Storage System Engineering and Applications: Applied Materials Development I
See more of this Group/Topical: Topical 8: Hydrogen Production and Storage

Storing Hydrogen Using (Ti,V)-Based Bcc Materials

Michel Latroche, Blaise Massicot and Jean-Marc Joubert, ICMPE UMR7182 CNRS, Thiais, France

Light vanadium-based alloys crystallizing in the body centred cubic (bcc) structure are promising materials for hydrogen storage. Adding titanium to vanadium improves the kinetics of the hydrogen absorption whereas addition of 3d metals such as iron enhances the reversible absorption capacity [1] and the activation process. The aim of this study is to highlight the influence of alloy composition on the hydrogenation properties (absorption pressure and capacity) of the bcc phase. The isothermal sections at 1000 °C and 1200 °C of the Ti–V–Fe system have been experimentally studied to get a better understanding of the bcc-region boundaries and equilibrium phases. For iron contents below 15 at.%, the isotherm at 1000 °C shows that the bcc phase is the only stable one. Moreover, our results indicate that the ternary extension of the $TiFe_2$ phase with the C14 structure reaches the composition $Ti_{33}V_{33}Fe_{33}$ [2]. The pressure-composition absorption isotherms of several bcc-alloys of the Ti–V–Fe system have been measured at various temperatures. For a constant Ti/V ratio, addition of iron increases the plateau pressure, as well as addition of vanadium for a given Ti/Fe ratio. The dependence of the capacity on the alloy composition has been determined [3] and will be presented as a function of the alloy composition (Fig. 1).

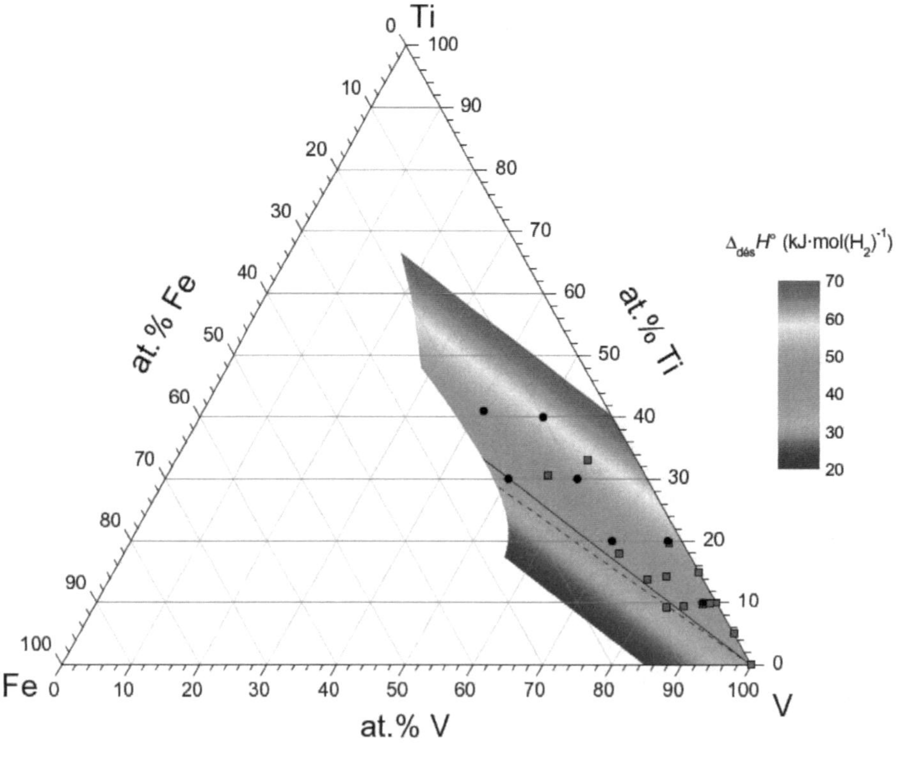

Figure 1 – Map of the enthalpy of dissociation for the dihydride of the Ti–V–Fe bcc system as a function of the composition. The dotted line corresponds to a constant lattice parameter whereas the full line stands for constant enthalpy.

[1] S. Challet, M. Latroche, and F. Heurtaux, J. Alloys Compd., 439, 294, 2007.
[2] B. Massicot, J.-M. Joubert and M. Latroche, Int. J. Mater. Res. (2010), doi: 10.3139/146.110379.
[3] B. Massicot, M. Latroche, J.-M. Joubert, Journal of Alloys and Compounds (2010), doi:10.1016/j.jallcom.2010.09.030.

(283a) Overview of Los Alamos National Laboratory's Engineering Research Efforts for On-Board Hydrogen Release From Chemical Hydrides

Troy Semelsberger, Los Alamos National Laboratory, Los Alamos, NM

An overview of Los Alamos National Laboratory's (LANL) research results in support of the DOE Hydrogen Storage Engineering Center of Excellence (HSECoE) will be presented. The purpose of the HSECoE is to design and implement novel engineering strategies for the on-board release of hydrogen from hydrogen storage media (i.e., metal hydrides, chemical hydrides, and sorbents). Los Alamos National Laboratory's research scope within the HSECoE includes fuel gauge sensor development, shelf-life modeling of hydrogen storage media, kinetics analyses and rate model development of chemical hydrides, designing and modeling novel chemical hydride reactors, and impurity identification, scrubbing and mitigation. Recent LANL results to be presented include: fuel gauge sensor development–demonstrated novel fuel gauge sensor with metal hydrides; shelf-life modeling–data collected from DSC and TGA were modeled to accurately predict thermal aging characteristics of candidate chemical hydrides (solid and liquid ammonia borane formulations); kinetics analyses and rate modeling–data collected from kinetics analyses were regressed into rate models that accurately predict reactor performance with liquid ammonia borane formulations; novel chemical hydride reactors–novel reactor schemes have been developed for the startup and transient operation using liquid ammonia borane formulations; impurity identification, scrubbing, and mitigation–ammonia, borazine, and diborane generated from AB have been identified and quantified using TGA-IR-MS, and can be scrubbed using various commercially available adsorbents. Impurity mitigation strategies include additives and accurate temperature control. In addition, the minimum impurity levels of borazine and diborane for the safe and effective operation of a fuel cell will also be presented.

See more of this Session: Hydrogen Storage System Engineeering and Applications: Applied Materials Development II
See more of this Group/Topical: Topical 8: Hydrogen Production and Storage

(283b) Chemical Hydrogen Storage Center of Excellence

Kevin C. Ott, Los Alamos National Laboratory, Los Alamos, NM

Recent results from DOE's Chemical Hydrogen Storage Center of Excellence will be presented. This Center is led by LANL and is one of DOE's three materials development Centers. The Center is a partnership among two national laboratories, 7 academic institutions, and two companies. This Center's work focuses on developing efficient processes that release hydrogen from covalent compounds and the subsequent efficient chemical regeneration of the spent fuels. Much of our work is directed at releasing hydrogen from amine-borane compounds such as ammonia borane and related derivatives. A high level overview of the chemistries we have explored for releasing hydrogen from these compounds will be presented, as well as a discussion of preliminary systems engineering issues and results. Regeneration of spent fuel is the Achilles heel of chemical hydrogen storage, and recent breakthrough results on the chemical regeneration of spent fuel and baseline process engineering and cost estimates will be discussed.

See more of this Session: Hydrogen Storage System Engineeering and Applications: Applied Materials Development II
See more of this Group/Topical: Topical 8: Hydrogen Production and Storage

Dehydrogenation Kinetics and Catalysis for Hydrogen Recovery from Organic Heteroaromatic Compounds for Hydrogen Storage

Farnaz Sotoodeh[a], *Benjamin Huber*[b], *Kevin J. Smith*[a,*]

[a]*Department of Chemical and Biological Engineering, University of British Columbia, Vancouver, BC Canada*

[b]*Chair I for Technical Chemistry, Technical University of Munich, Garching, Germany*

**Corresponding author: +1-604-822-6003, kjs@interchange.ubc.ca*

Abstract

The complete recovery of H_2 from dodecahydro-N-ethylcarbazole was achieved at 443 K and 101 kPa using Pd catalysts prepared by wet impregnation with calcination in He rather than air. The reaction proceeded to complete conversion within 22 min and complete H_2 recovery (5.8 wt %) within 1.6 h over a 4 wt % Pd/SiO_2 catalyst. The dehydrogenation of dodecahydro-N-ethylcarbazole was structure sensitive. Decreasing the Pd particle size to approximately 9 nm increased the turn-over frequency (TOF) and selectivity to the completely dehydrogenated product. The dehydrogenation rate of dodecahydro-N-ethylcarbazole was compared to that of dodecahydrocarbazole and dodecahydrofluorene to determine the effect of the N heteroatom and the ethyl group of dodecahydro-N-ethylcarbazole. The lower TOF for dodecahydrocarbazole was attributed to a strong adsorption of the dehydrogenated products to Pd through the N atom, whereas the ethyl group in dodecahydro-N-ethylcarbazole prevented a strong N interaction with the surface. DFT results showed that dodecahydrocarbazole and dodecahydrofluorene were adsorbed on Pd more strongly than dodecahydro-N-ethylcarbazole, leading to a significant decrease in their TOFs for H_2 recovery.

Keywords: Hydrogen storage, dehydrogenation, catalyst, kinetics, DFT, dodecahydro-N-ethylcarbazole, dodecahydrocarbazole, dodecahydrofluorene.

1. INTRODUCTION

The use of hydrogen as an alternative, benign energy carrier is limited by the technology gap that exits for H_2 storage. Compressed or liquefied hydrogen is costly and the gravimetric storage capacity is low. A more practical option is off-board regenerable materials such as organic heteroaromatics on which hydrogen is stored by chemical reaction [1]. Compounds such as carbazole and N-ethylcarbazole, with gravimetric storage capacities > 5.5 wt % are potential candidates. Their hydrogenated forms can release hydrogen at low temperatures (< 473 K) [2,3,4,5] and can be used for on-board hydrogen recovery applications. However, dehydrogenation reactions of the aromatic compounds have been reported to be very slow. In previous work [4,5], 100 % conversion of dodecahydro-N-ethylcarbazole (with no selectivity to the completely dehydrogenated product N-ethylcarbazole) and 53 % conversion of dodecahydrocarbazole was obtained after 17 h at 443 K and 101 kPa over a 5 wt % Pd/SiO_2 catalyst with a Pd particle size of 25 nm. In work by Wang et al. [6] iridium complexes were shown to be active catalysts for the dehydrogenation of dodecahydro-N-ethyl carbazole at 473 K. However, dehydrogenation to the fully unsaturated N-ethylcarbazole did not occur in most cases and octahydro-N-ethylcarbazole and tetrahydro-N-ethylcarbazole were the only products. Also, catalyst activity was much lower at 423 K during 48 h reaction time. Dehydrogenation of tetrahydrocarbazole at 413 K was reported to proceed to 81 % conversion over a 5 wt % Pd/Al_2O_3 catalyst in 27 h [7,8]. Despite a 100 % selectivity to carbazole, product inhibition decreased the dehydrogenation rate significantly. Dehydrogenation of decalin using a Pt/C catalyst and a batch-type membrane reactor at high temperatures (533-593 K) resulted in conversions below 85 % after 2.2 h [9] and 95 % conversion within 1.5 h at 553 K [10]. However, the reported temperature was too high and impractical for hydrogen recovery for transport applications. Despite the available literature data, there are limited studies on the effect of catalyst particle size on the hydrogenation/dehydrogenation of aromatic compounds. In one study the hydrogenation of 2,4-dinitrotoluene over Pd/SiO_2 was favored by increasing the Pd particle size [11]. However, in another study, the hydrogenation of 2,4-dinitrotoluene over Pd/C catalysts was found to be a structure insensitive reaction [12] where the increases of the specific activity on the larger metal particles was attributed to diffusion limitations within the pores of the microporous carbon used. Crawford et al. [7] investigated the mechanism of tetrahydrocarbazole dehydrogenation over a Pd catalyst, and showed by experiment and DFT calculations, that the reaction proceeded through a flat tetrahydrocarbazole adsorption geometry involving several Pd surface sites, indicative of a structure sensitive reaction. In the present work, using a series of Pd catalysts with different Pd particle sizes, structure sensitivity of the dehydrogenation of dodecahydro-N-ethylcarbazole was studied. The results were also compared with the dehydrogenation reactions of dodecahydrocarbazole and dodecahydrofluorene. DFT calculations were used to optimize and compare the adsorption geometry of each molecule to the Pd surface in order to explain the difference in the observed reaction rates.

2. EXPERIMENTAL

2-1 Catalyst Preparation and characterization

Silica supported Pd catalysts with Pd loading of 0.5 wt % to 10 wt %, were prepared by the incipient wetness impregnation of silica gel (Sigma–Aldrich). The required amount of an aqueous solution of HCl and $PdCl_2$ (Sigma–Aldrich, 99.9+ %) precursor were added dropwise to the silica gel and left to age for 48 h. The samples were dried at 393 K for 8 h and calcined at 773 K for 6 h in a 30 cm^3 (STP)/min flow of He. The catalysts were then reduced in 30 cm^3

(STP)/min of H_2 by heating 10 K/min to 673 K and holding the final temperature 1 h before cooling to room temperature in a 30 cm^3 (STP)/min flow of He. A commercial 5 wt % Pd on activated carbon catalyst (Aldrich, 5 wt % Pd, reduced) was used for comparative dehydrogenation reactions of dodecahydrocarbazole and dodecahydrofluorene.

Temperature-programmed desorption (TPD) of CO was performed using a Micromeritics Autochem II 2920 analyzer. About 0.2 g of the dried catalyst was calcined at 773 K for 6 h in a flow of Ar at 30 cm^3 (STP)/min and reduced at 673 K in a 30 cm^3 (STP)/min flow of 10 % H_2 in Ar for 1 h. After flushing the catalyst in 30 cm^3 (STP)/min flow of Ar for 1 h, it was exposed to 50 cm^3 (STP)/min flow of 10 % CO in He at 313 K for 1 h and flushed with 50 cm^3 (STP)/min He for 2 h. The TPD was then started by heating the sample to 773 K at a rate of 10 K/min and holding the final temperature for 1 h. The catalysts were also characterized by scanning electron microscopy (SEM) and energy dispersive X-ray analysis (EDX). SEM, EDX and EDX-mapping were done using a variable pressure 120 keV Hitachi S-3000N equipped with a light element EDX detector. A FEI Tecnai G2 200kV with a LaB_6 filament capable of 1.4 Å point-to-point resolution was used for high resolution transmission electron microscopy (HRTEM). Pd/SiO_2 samples with variable Pd loadings were ground to a fine powder using an agate mortar and pestle and dispersed in ethanol ultrasonically. A droplet of the suspension was then placed on a 200 mesh copper grid coated with formvar carbon and was left to dry before analysis.

2-2 Catalyst activity

The dehydrogenation reactions were carried out at 443 K and 101 kPa in a 50 cm^3 glass flask reactor using the reduced Pd/SiO_2 catalysts and the product obtained from N-ethylcarbazole hydrogenation [4,5]. The reactor was stirred and heated to the desired temperature with helium used as a carrier gas for continuous removal of the produced hydrogen. The exit gas composition was continuously monitored using a quadrupole mass spectrometer (SRC Residual Gas Analyzer, 200 amu). Liquid samples were also withdrawn from the reactor periodically. The analysis of the liquid samples from the dehydrogenation reactions was done by a Shimadzu QP-2010S GC/MS using a Restek RTX5 30 m × 0.25 mm capillary column. For comparison, the dehydrogenation reactions of dodecahydrocarbazole (obtained from carbazole hydrogenation using the same procedure as for N-ethylcarbazole hydrogenation) and dodecahydrofluorene (Aldrich, 97 %) were carried out in the same dehydrogenation reactor at temperatures from 423 K to 443 K and 101 kPa over the Pd/C catalyst.

3. CALCULATIONS

DFT calculations were performed using Material Studio 4.4.0.0. DMol^3 [13,14] to compare the adsorption geometry of dodecahydro-N-ethylcarbazole, dodecahydrocarbazole and dodecahydrofluorene. Double-numeric quality basis set (DNP) with the gradient-corrected GGA functional of Becke [15] and PBE description of exchange and correlation effects [16,17] were used in the calculations. For the numerical integration, Fermi smearing of 2×10^{-3} Ha, real-space cut-off of 4.0 Å and MEDIUM quality mesh size were considered. The Pd(111) surface was modeled by periodic slabs of four Pd layers with a 30 Å of vacuum region between the slabs. The molecule and the top layer of the Pd(111) surface were allowed to relax. A (3×3) supercell was chosen to model the adsorption of each molecule on the Pd(111) surface. Density functional semi-core pseudopotentials (DSPP) were used for the Pd substrate. The tolerances of energy, gradient, and displacement convergence were 2×10^{-5} Ha, 4×10^{-3} Ha/Å, and 5×10^{-3} Å, respectively. Adsorption energies were determined as $E_{adsorption} = E_{adsorbed\ molecule\ /\ surface} - E_{molecule\ in\ vacuum} - E_{surface}$, where $E_{adsorbed\ molecule\ /\ surface}$ is the energy of the adsorption system, $E_{molecule\ in\ vacuum}$ is the energy of the adsorbed molecule in vacuum, and $E_{surface}$ is the energy of the surface.

4. RESULTS AND DISCUSSION

4-1 Catalyst properties

The CO uptake results from the TPD experiments were used to estimate the catalyst dispersion at each Pd loading, as well as the Pd particle diameter. The results are summarized in **Table 1**. The 5 wt% Pd/SiO_2 data show that calcination in He rather than air resulted in an almost twofold increase in the Pd metal dispersion. Selected HRTEM results shown in Figure 1 for the 4 wt % Pd/SiO_2 sample calcined in He and reduced in H_2 were also in good agreement with the Pd particle size obtained from the CO TPD experiments.

Table 1. Catalyst properties obtained from CO TPD and HRTEM size analysis for Pd/SiO_2 catalyst with Pd loading of 0.5-10 wt %.

Pd loading (wt%)	CO desorbed (μmol/g)	Dispersion (%)	Pd particle size[b] (nm)	Average Pd particle diameter[c] (nm)
0.5	20.99	52	1.9	3.7 ± 0.25[d]
0.7	25.29	45	2.2	
1	22.48	28	3.6	
2	25.66	16	6.3	
3	31.03	13	7.7	
4	34.36	11	9.1	7.3 ± 0.49[d]
5	30.34	8	12.5	
10	53.10	6.5	15.4	17.1 ± 0.97[d]
5[a]	-----	4.2	23.8	

[a] Prepared by calcination in air [4,5]. [b] Obtained from TPD analysis. [c] Calculated by taking the average of particle diameters obtained from HRTEM imaging. [d] Standard error of mean.

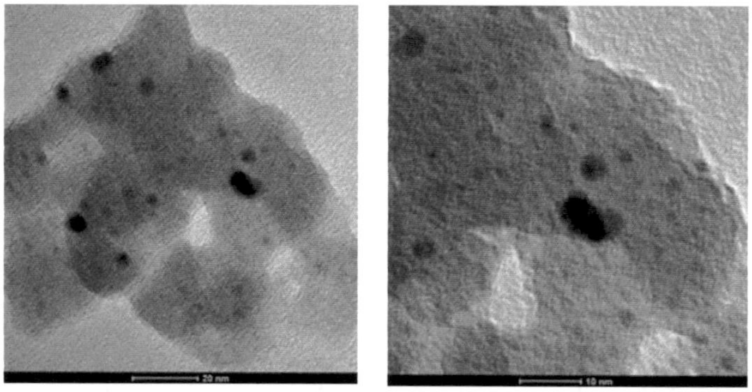

Figure 1. Selected HRTEM images of reduced 4 wt % Pd/SiO_2 sample calcined in He flow at 500 °C for 6 h prior to reduction in H_2 at 400 °C for 1 h; scale bar (left) 20 nm, (right) 10 nm.

From the SEM and EDX elemental mapping shown in Figure 2 for the reduced 5 wt % Pd/SiO_2 it can be seen that almost no Cl was identified after reducing the sample in H_2 flow at 400 °C for 1 h. The results confirm the removal of the Cl in the catalyst calcined in He after the thermal treatments.

Figure 2. Selected SEM micrographs and elemental mapping of the reduced 5 wt % Pd/SiO_2 sample, calcined in He at 500 °C for 6 h.

4-2 Catalyst activity

Figure 3 shows the product distribution data over time obtained from the liquid sample analysis of the dehydrogenation reactions at 443 K and 101 kPa over the 4 wt % Pd/SiO_2. Octahydro-N-ethylcarbazole and tetrahydro-N-ethylcarbazole were the primary products and the reactions proceeded with further conversion of these intermediates and the production of the completely dehydrogenated product, N-ethylcarbazole.

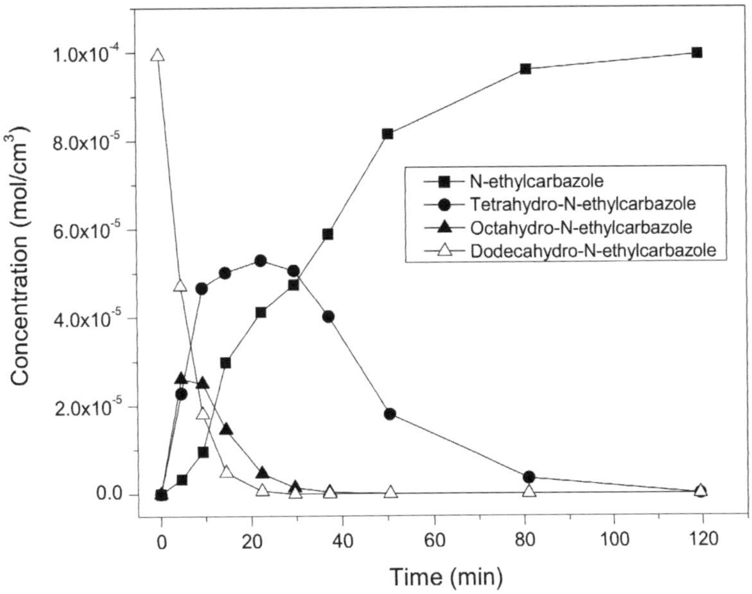

Figure 3. Product distribution for dodecahydro-N-ethylcarbazole dehydrogenation at 443 K and 101 kPa over 4 wt % Pd/SiO_2 (Pd particle diameter~9 nm, metal dispersion~11%).

First-order kinetics were fitted to the concentration of the reactant, dodecahydro-N-ethylcarbazole, to obtain the reaction rate constants shown in Table 2. TOFs per Pd surface atom with respect to the consumption of dodecahydro-N-ethylcarbazole were calculated using the obtained first-order rate constants and the initial concentration of dodecahydro-N-ethylcarbazole.

The reaction rate constant per gram of catalyst increased from 0.0036 $min^{-1}.g_{cat}^{-1}$.lit to 0.0921 $min^{-1}.g_{cat}^{-1}$.lit by increasing the Pd loading from 0.5 wt % to 4 wt %, and decreased to 0.0618 $min^{-1}.g_{cat}^{-1}$.lit by further increase in the Pd loading to 10 wt %. The same trend was observed for the TOF of the reaction, where the 4 wt % catalyst resulted in the highest TOF of 221.2 min^{-1} for dodecahydro-N-ethylcarbazole dehydrogenation. Complete H_2 recovery was obtained in only 1.6 h over the 4 wt % Pd catalyst.

Table 2. Kinetic results and H_2 recovery obtained for dehydrogenation of dodecahydro-N-ethylcarbazole at 443 K and 101 kPa.

Pd loading (wt %)	Rate constant[b] ($min^{-1}.g_{cat}^{-1}$.lit)	Initial concentration (mol. $cm^{3\ -1}$)	TOF[c] (min^{-1})	Final H_2 recovery[d] (wt %)
0.5	0.0036	9.93×10^{-5}	14.6	3.8 (6 h)[e]
0.7	0.0049	9.93×10^{-5}	16.3	4.4 (6 h)
1	0.0134	9.93×10^{-5}	50.4	5.2 (6 h)
2	0.0296	9.93×10^{-5}	97.8	5.8 (2.7 h)
3	0.0427	9.93×10^{-5}	115.7	5.8 (1.9 h)
4	0.0921	9.93×10^{-5}	221.2	5.8 (1.6 h)
5	0.0593	9.88×10^{-5}	155.7	5.8 (3 h)
10	0.0618	9.93×10^{-5}	100.5	5.8 (6 h)
5[a]	0.0045	1.40×10^{-4}	31.8	4.0 (17 h)

[a] Reported by Sotoodeh et al. [4]. [b] With respect to dodecahydro-N-ethylcarbazole consumption. [c] Molecule dodecahydro-N-ethylcarbazole. min^{-1}. Pd surface $atom^{-1}$. [d] Theoretical amount of released H_2 in grams divided by the weight of hydrogenated compound in grams (theoretical H_2 recovery capacity for doecahydro-N-ethylcarbazole is 5.8 wt %). [e] Experimental time for the obtained H_2 recovery.

The reported results are significant for hydrogen recovery applications, as the dehydrogenation reactions have been reported to be very slow with no selectivity to the completely dehydrogenated product, N-ethylcarbazole, using 5 wt % Pd/SiO_2 [4,5] as well as iridium complexes [6]. Previous reports on the dehydrogenation of tetrahydrocarbazole at 413 K over a 5 wt % Pd/Al_2O_3 [7,8], reported that tetrahydrocarbazole adsorbed almost parallel to the surface on several Pd atoms, and dehydrogenation was reported to proceed to 81 % conversion in 27 h at 413 K and 101 kPa. The product inhibition by carbazole was shown to slow the reaction rate significantly. Decalin was also dehydrogenated over a Pt/C catalyst at temperatures of 533-593 K and conversions were below 85 % in 2.2 h [9]. The reported reaction temperature however, was too high and impractical for hydrogen recovery for transport applications. In the present work, the change in dehydrogenation selectivity to N-ethylcarbazole with a change in Pd particle size shows that the dehydrogenation reaction of dodecahydro-N-ethylcarbazole is structure dependent and that the particle size plays an important role in providing an adequate number of catalytic sites for the adsorption of the dodecahydro-N-ethylcarbazole molecule and the other dehydrogenation intermediates to the surface.

Dodecahydro-N-ethylcarbazole dehydrogenation was compared to dodecahydrocarbazole and dodecahydrofluorene dehydrogenation with H_2 capacities of 6.70 and 6.74 wt %, respectively. Table 3 summarizes the obtained kinetic results as well as the H_2 recovery for the dehydrogenation reactions of dodecahydro-N-ethylcarbazole, dodecahydrocarbazole and dodecahydrofluorene. The reaction TOFs obtained for dodecahydro-N-ethylcarbazole over the catalysts with 24 and 6 nm particle size were significantly higher than those of dodecahydrocarbazole. Furthermore, decreasing the Pd particle size from 24 nm to about 5-6 nm resulted in a dramatic increase in the reaction TOF for both reactants, pointing to the structure sensitivity of the dehydrogenation reactions. The lower reaction TOF for dodecahydrocarbazole dehydrogenation compared to dodecahydro-N-ethylcarbazole is attributed to the effect of the N heteroatom lone pair which interact with the catalyst surface and decrease the reaction rate and subsequently the H_2 recovery. Catalyst poisoning by the product carbazole was also reported for the dehydrogenation reaction of tetrahydrocarbazole [8]. The interaction of the N heteroatom with the catalyst surface, however, decreased for dodecahydro-N-ethylcarbazole and its dehydrogenated product, most likely due to the steric hindrance of the ethyl group. Comparing the kinetics of the dehydrogenation of dodecahydrofluorene with dodecahydrocarbazole, on the other hand (Table 3), it is seen that a much lower TOF for dodecahydrofluorene was obtained compared to dodecahydrocarbazole. Furthermore, the very low H_2 recovery at the reaction temperature of 443 K after an extended reaction period, confirms that the N heteroatom improves the dehydrogenation of organic heteroaromatics significantly [18].

Table 3. Comparison of dehydrogenation reactions of dodecahydro-N-ethylcarbazole, dodecahydrocarbazole and dodecahydrofluorene at 443 K and 101 kPa.

Reactant	Catalyst particle size (nm)	TOF[a] (min^{-1})	Conversion (%)	H_2 recovery (wt %)
Dodecahydro-N-ethylcarbazole	24 (5wt% Pd/SiO_2)[b]	31.85	100 (17 h)[d]	4.00 (17 h)[d]
	6 (2wt% Pd/SiO_2)	97.78	100 (40 min)	5.79 (2.7 h)
Dodecahydrocarbazole	24 (5wt% Pd/SiO_2)[b]	0.79	53 (17 h)	1.90 (17 h)
	5 (5wt% Pd/C[c])	20.4	100 (30 min)	6.56 (6.3 h)
Dodecahydrofluorene	5 (5wt% Pd/C)	0.27	12.4 (> 20 h)	0.75 (70 h)

[a] Molecule of reactant. min^{-1}. Pd surface $atom^{-1}$. [b] Reported by Sotoodeh et. al. [4]. [c] Pd/C with Pd particle diameter ~5nm and metal dispersion~24% obtained from CO chemisorption experiment. [d] Experimental time.

4-3 Calculations

The effect of the N heteroatom on the dehydrogenation reaction was also studied by comparing the adsorption energies of dodecahydro-N-ethylcarbazole, dodecahydro-carbazole and dodecahydrofluorene on Pd, summarized in Table 4. The optimized adsorption geometries revealed that the adsorption of each molecule required several catalytic sites. Dodecahydrofluorene, showed a much stronger adsorption than dodecahydrocarbazole and dodecahydro-N-ethylcarbazole, as shown by a short $\sigma_{Pd\text{-}C}$ bond distance of 2.11 Å and an adsorption energy of 180.8 kJ/mol. The strong adsorption of dodecahydrofluorene likely causes the very slow dehydrogenation reaction rate of this compound.

Table 4. Comparison between the adsorbed structures and energies over Pd(111)

Compound	Dodecahydro-N-ethylcarbazole	Dodecahydrocarbazole	Dodecahydrofluorene
	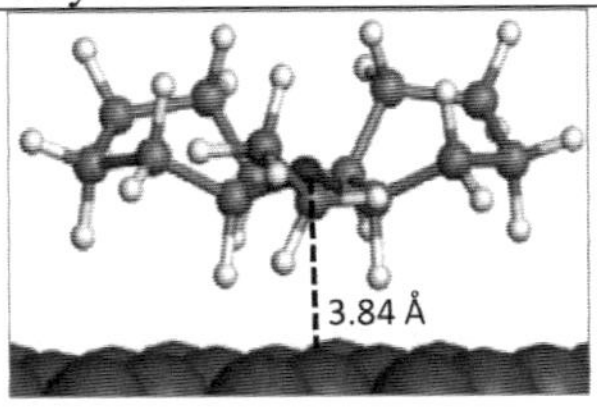	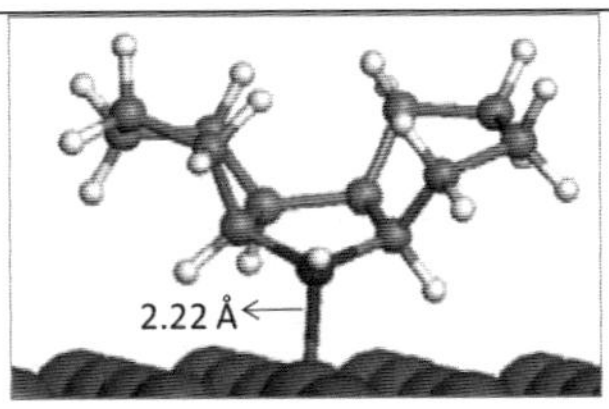	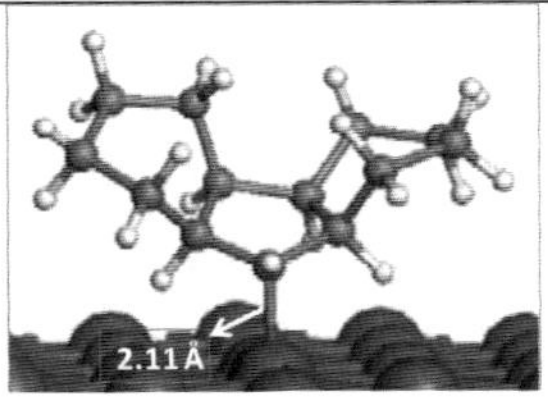
Adsorption energy, kJ/mol	-95.0	-109.4	-180.8

Dodecahydrocarbazole also had a stronger adsorption to the catalyst surface compared to dodecahydro-N-ethylcarbazole, which is likely due to the strong interaction of N heteroatom in this molecule with the catalyst surface.

5. CONCLUSION

Dodecahydro-N-ethylcarbhazole dehydrogenation was found to be structure sensitive. Pd catalysts prepared by calcination in He rather than air improved Pd dispersion, which led to complete H_2 recovery over a 4 wt % Pd/SiO_2 in less than 2 h at 443 K, a much higher catalyst activity compared to previously reported results. Although dodecahydrocarbazole and its dehydrogenated products were shown to poison the catalytic active sites significantly, due to the strong interaction of the N heteroatom lone pair with the surface, the presence of a N heteroatom in the polycyclic aromatic structures increases the H_2 recovery rate.

ACKNOWLEDGEMENT

Funding for the present study from Natural Sciences and Engineering Research Council of Canada is gratefully acknowledged. The authors also would like to acknowledge the Bio-Imaging Facility at the department of Botany of the University of British Columbia for the high resolution transmission electron microscopy (HRTEM) imaging.

Bibliography

[1] Petrovic J, Read C, Thomas G, Ordaz G. Satyapal S, *Catalysis Today*, vol. 120, pp. 246–256, 2007.

[2] G Pez, A Scott, AC Cooper, and H Cheng, US patents 7101530B2, 2006.

[3] U Eberle, M Felderhoff, and F Schth, *Angew. Chem. Int. Ed.*, vol. 48, pp. 6608–6630, 2009.

[4] F Sotoodeh, L Zhao, and KJ Smith, *Applied Catalysis A: General*, vol. 362, pp. 155-162, 2009.

[5] F Sotoodeh and KJ Smith, *Industrial and Engineering Chemistry Research*, vol. 49, pp. 1018–1026, 2010.

[6] Zh Wang, I Tonks, J Belli, and CM Jensen, *Organomet. Chem.*

[7] P Crawford et al., *J. Phys. Chem.*, vol. 111, pp. 6434–6439, 2007.

[8] KT Hindle et al., *J. of Catal.*, vol. 251, pp. 338–344, 2007.

[9] RO Loutfy and EM Veksler, in *the International Hydrogen Energy Forum*, Munich, Germany, Sept 11-15, 2000, pp. 335-340.

[10] Sh Hodoshima, H Arai, Sh Takaiwa, and Y Saito, *Intl. J. Hydrogen Energy*, vol. 28, pp. 1255–1262, 2003.

[11] A Benedetti et al., *Catal. Lett.*, vol. 10, p. 2115, 1991.

[12] DJ Suh, TJ Park, and S Ihm, *Ind. Eng. Chem. Res.*, vol. 31, p. 1849, 1992.

[13] M Ohara, Y Kim, S Yanagisawa, Y Morikawa, and M Kawai, *Phys. Rev. Lett.*, vol. 136104, p. 100, 2008.

[14] W Dou, *J. of Chem. Phys.*, vol. 244711, p. 128, 2008.

[15] AD Becke, *Phys. Rev. A*, vol. 38, p. 3098, 1988.

[16] B Hammer, LB Hansen, and JK Norskov, *Phys. Rev. B*, vol. 59, p. 7413, 1999.

[17] JP Perdew, K Burke, and M Ernzerhof, *Phys. Rev. Lett.*, vol. 77, p. 3865, 1996.

[18] Y Cui et al., *New J Chem*, vol. 32, pp. 1027-1037, 2008.

Dynamic Modeling and Simulation of Ammonia Borane Hydrogen Storage Systems

Ewa Rönnebro, Maruthi Devarakonda, Kriston Brooks, Scot Rassat and Darrell Herling
Pacific Northwest National Laboratory, Richland, WA 99352, USA

Out of the materials available for chemical hydrogen storage in PEM fuel cell applications, ammonia borane (AB, NH_3BH_3) has a high hydrogen storage capacity (up to 19.6% by weight for the release of three hydrogen molecules). Therefore, AB was chosen in our chemical hydride simulation studies. A new systems concept based on augers, ballast tank and reactor, hydrogen heat exchanger and H_2 burner was designed and implemented in simulation. One dimensional models based on conservation of mass, species and energy were used to predict important state variables such as reactant and product concentrations, temperatures of various components, flow rates, along with pressure, in various components of the storage system. Control signals to various components are governed by a control system which is modeled as an independent subsystem. Various subsystem components in the models were coded as C language S-functions and implemented in Matlab/Simulink environment. System simulation results for vehicle start-up and transient drive cycle cases are presented. In the following, we will describe the materials operating requirements and modeling of ammonia borane.

Materials Operating Requirements for Chemical Hydrides

The US Department of Energy (DOE) Hydrogen Storage Engineering Center of Excellence (HSECoE) team is working collaboratively to develop an integrated system model combining hydrogen storage, fuel cell and vehicle performance sub-models[1] based on materials operating requirements. The overall objective of the HSECoE is to model and optimize the necessary hardware required to build subscale prototype hydrogen storage systems for light duty vehicle applications. The HSECoE is building on the materials discovery and development results of each of the three DOE Materials Center of Excellences (MCoE) and their acquired materials' data, i.e. from the three different groups of metal hydrides, chemical hydrides and adsorbents.

During the first year of the HSECoE (FY09), materials operating requirements were established for the three different groups of materials, a materials screening methodology was identified and materials candidates selected with the input of the existing MCoEs. Relevant data has been compiled from the center where available and from the literature, into a data base for HSECoE use. The selected materials were grouped into "down-selected" and "up-selected" materials. We define *Tier 1 and Tier 2* up-selected materials. For *Tier 1* "Developed Materials" system analysis is being performed on up-selected candidates and necessary engineering properties measured of candidate materials with the greatest potential for meeting the DOE 2015 targets. *Tier 2* "Developing Materials" are up-selected materials under performance evaluation and materials properties collected and measured if necessary. Down-selected materials are those materials found to not improve system performance relative to up-selected materials, and thus not for further consideration.

In order for a material to be considered for application within the HSECoE, it must first pass minimum screening criteria established for each materials group which will provide a rough assessment of its capabilities and potential to meet the 2010 DOE targets. This requires minimal but strategic data and involves hydrogen content, crystal density, enthalpy, operation temperatures, kinetics and cost. For chemical hydrides, the gravimetric capacity should be >9wt% materials capacity, hydrogen desorption <150°C and hydrogen release rate >30g/s. Moreover, the enthalpy of formation needs to be <20kJ/mol and the crystal density >1g/cm^3. The HSECoE has up-selected three chemical hydride materials for further considerations; ammonia borane (solid and liquid), AlH_3 and $LiAlH_4$. Once past the screen, a full data set must be generated covering the materials performance characteristics over the temperature and pressure range of use, in order to begin detailed modeling assessment.

Out of the materials available for chemical hydrogen storage for off-board regenerable applications, ammonia borane (AB, NH_3BH_3) is a promising material for meeting the DOE targets due to an inherently large gravimetric and volumetric hydrogen density of 19.6wt% hydrogen and 0.145 $kgL^{-1}H_2$. Therefore, the HSECoE up-selected this material as a '*Tier 1* Developed material' for chemical hydride simulation studies. Unlike the reversible hydrides which have endothermic H-release, AB undergoes an exothermic H-release mechanism. It thermally decomposes between 70 and 110°C to polyaminoborane (PAB), $[NH_2BH_2]_n$ and hydrogen. Thereafter, PAB decomposes over a broad temperature range from 110°C to 200°C to from polyiminoborane (PIB), $[NHBH]_n$, hydrogen and a small fraction of borazine, $[N_3B_3H_6]$. This is in total 2.5 equivalents of hydrogen, or ca 12-13wt%. If heating to temperatures above 500°C, PIB is decomposed to release the remaining hydrogen upon formation of BN, however, BN is too stable for practical applications, thus, 2.5 equivalents is the realistic hydrogen release.

Several fundamental studies on AB have been performed at PNNL by Autrey and co-workers for the DOE Chemical Hydride CoE to understand the thermal stability and chemical kinetics of hydrogen release from AB[2-5]. Aardahl and Rassat recently reviewed chemical hydrogen storage systems and discussed system engineering considerations[4]. They pointed out that chemical storage is different from other storage solutions since the fuel must be moved in addition to releasing and transporting the hydrogen. To reduce thermal management there is a need for a hydrogen release reactor to be separated from the fuel storage. This is also important to be able to control the hydrogen release rate. Engineering a reaction with exothermal hydrogen release requires a different approach compared to the endothermal hydrogen release reactions. It is important to manage the heat evolved during hydrogen release and as will be discussed in more detail below, this can be done by using a radiator. Other components involve burners and heat exchangers.

Several reports on modeling hydrogen storage systems are found in the literature since Lucas and Richards presented their pioneering work on modeling hydride beds in 1984[6]. However, a large majority of the models were focused on metal hydride storage systems[7,8]. This document presents a chemical hydride storage model. Chemical hydrides are not regenerated on-board and one effect of this is that the materials need to be moved within the system. It's important to consider whether moving the material as a solid material, slurry or a liquid. The current model is based on solid ammonia borane.

Modeling Chemical Hydride, Hydrogen Storage Systems

This section presents an overview of chemical hydride, hydrogen storage system modeling activities at PNNL. For the first time in the literature, a systems model for an ammonia bead reactor system was recently presented[9]. The main components in the bead reactor system were the augers (hot and cold), ballast tank and reactor, radiator and H_2 burner. A baseline control system varies the flow rate of AB into the hot auger based on the ballast tank states (temperature and pressure) and fuel cell demand. All the model equations are not presented here for conciseness, instead key features such as AB kinetics and the ballast tank dynamics are discussed here.

Based on previous investigations of H_2 release from AB under isothermal conditions, a simple kinetic model at temperatures greater than ~110°C was developed. The AB conversion plots as a function of time and at a given temperature, similar to the one shown in Figure 1, were sigmoidal, characteristic of nucleation and growth phenomena in solids.

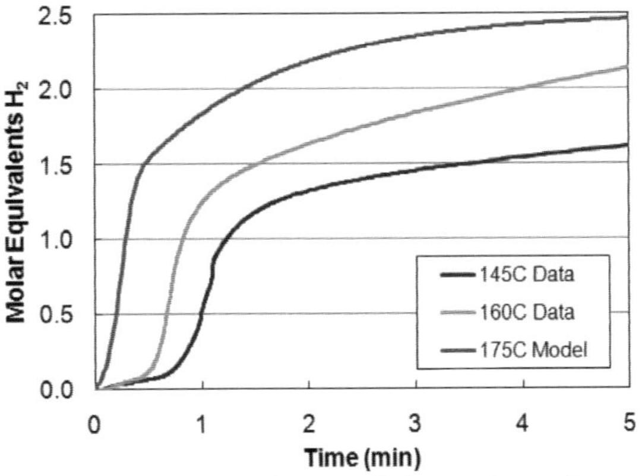

Figure 1: Predicted H_2 release from AB based on three global models

A linear combination of Avrami kinetic expressions, as shown in E1, was used to model the H_2 release from AB.

$$\chi = 1.1\{(1-\exp(-(k_1 t^*)^3))\} + 1.4\{(1-\exp(-(k_2 t^*)))\} \qquad \text{(E1)}$$

Here t^* represents the reaction time under isothermal conditions to attain a conversion fraction 'χ' (moles of H_2 released per mole of AB). The rate constants k_i are defined using Arrhenius law. The values of the rate parameters are given in [9]. The reaction rate (R_{rxn}) is obtained by taking the derivative of E1 and is shown in E2.

$$R_{rxn} = 3.3k_1^3 t^{*2} \exp(-(k_1 t^*)^3) + 1.4k_2 \exp(-(k_2 t^*)) \qquad \text{(E2)}$$

The effective reaction time (t^*) is calculated by solving E1 for a local AB conversion (χ) and temperature (T_{AB}) and is implemented in the Matlab/Simulink simulation as a 2D look-up table. The AB conversion as a function of time and location along the ballast tank is calculated as shown in E3.

$$\frac{\partial \chi}{\partial t} = -u_{AB}\frac{\partial \chi}{\partial x} + R_{rxn} \qquad \text{(E3)}$$

u_{AB} is the velocity of AB in the tank calculated as a function of AB flow rate, bulk density and area of the tank. Detailed information on the model was presented in our previous publication[9]. The temperature of AB in the tank is given by E4.

$$\rho_{AB,bulk} C_{p,AB} \frac{\partial T_{AB}}{\partial t} = \frac{\Delta H_{rxn} R_{rxn} \rho_{AB,bulk}}{MW_{AB}} - \frac{C_{p,AB} \dot{m}_{AB,in}}{A} \frac{\partial T_{AB}}{\partial x} \quad \text{(E4)}$$

The feed temperature at the tank inlet is the temperature exiting the hot auger. If $\chi < 1$, ΔH_{rxn} = 22 kJ/mol H_2 and if $\chi > 1$, ΔH_{rxn} = 15 kJ/mol H_2. All the component models are coded as C language S-functions and are implemented in Matlab/Simulink environment. A variable step solver 'ode23tb' is used to address the numerical stiffness associated with kinetic equations[10].

Model Based Control of Chemical Hydride, Hydrogen Storage Systems

For operating conditions with T_{AB} > 200°C (i.e., after cold-start), the baseline controller is defined as shown in E5.

$$\dot{m}_{AB} = \frac{FC_{demand} MW_{AB}}{\chi P_{BT}} P_{set} \quad \text{(E5)}$$

Here, FC_{demand} is the hydrogen demand from the fuel cell (mol/s), P_{BT} is the pressure of AB in the ballast tank (atmospheres), and P_{set} is the nominal set-point of the tank pressure (chosen as 500 atmospheres). For the case presented, a constant flow of 0.2 mol/s of H_2 is used to fire the burner when the temperature of the incoming oil is less than 80°C. To demonstrate the baseline controller in simulation, fuel cell H_2 demand data from US06 drive cycle for a non-hybrid case (Courtesy: NREL[11]) is fed as input to the simulation and the system response (temperatures and pressure) is shown in Figure 2.

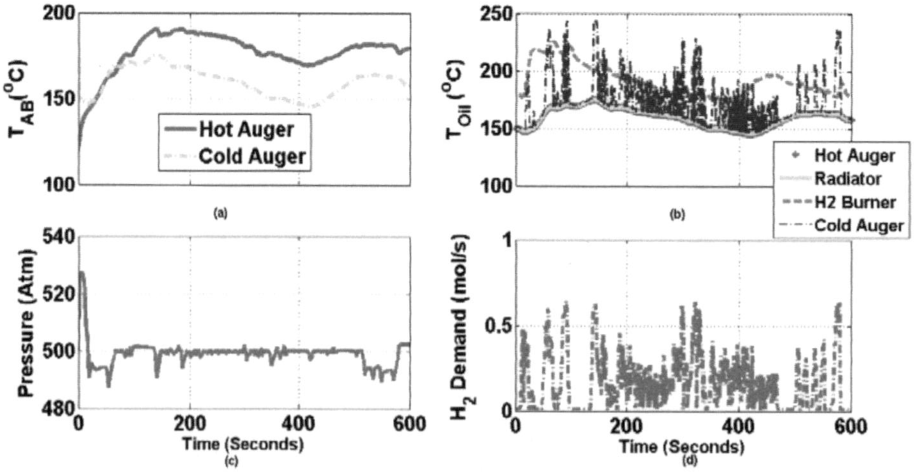

Figure 2: Storage system response for H_2 demand from the fuel cell during a transient cycle – (a) AB Temperature; (b) Oil Temperature; (c) Tank Pressure; (d) Hydrogen demand from fuel cell

In addition to the non-hybrid case, transient simulations (from US06 drive cycle) using the fuel cell H_2 demand data for a 20% hybrid case were also conducted. The integral output of H_2 from the reactor system (referred to as HSS Response) for both the cases is calculated and is compared to the integral demand from the fuel cell as shown in Figure 3. In addition, a detailed report on system dynamics and control of the reactor system under study was presented by Devarakonda et al[12].

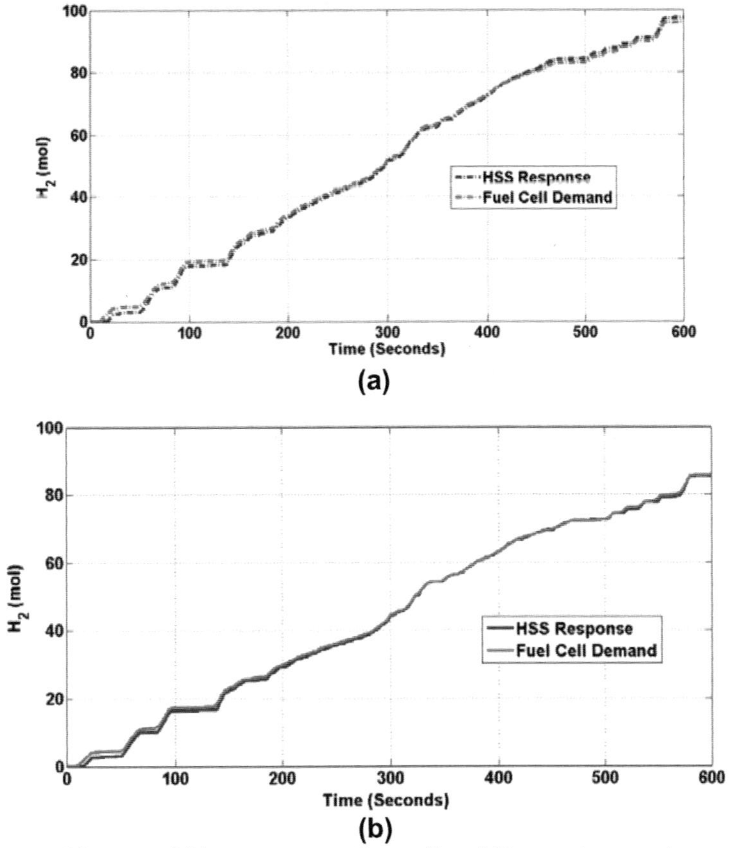

Figure 3: Comparison of integral H_2 response from the AB reactor system and fuel cell demand for (a) 0% hybrid and (b) 20% hybrid cases

Conclusions

We selected ammonia borane, NH_3BH_3, for modeling and simulation of a chemical hydrogen storage bed based on its promising materials operating requirements and potential to meet the US DOE system targets. A representative model for an AB based bead reactor system was developed for the first time and was successfully implemented in a simulation platform. Energy balance and chemical kinetics equations yielded the temperatures and product concentrations across various components of the reactor system. Molar conversion of AB into H_2 was implemented through Avrami kinetics and was used in the ballast tank and reactor model. A baseline controller was developed using the ballast tank states and hydrogen demand from the fuel cell to determine the flow rate of AB in the system. System simulation results show appropriate trends in reactor system dynamics, while controls analysis demonstrated the usefulness of the proposed controller form. Future work includes the implementation of an AB slurry reactor system in simulation.

Acknowledgements

The authors acknowledge support from the U.S. Department of Energy's Office of Energy Efficiency and Renewable Energy. This work was performed as part of Hydrogen Storage Engineering Center of Excellence. Pacific Northwest National Laboratory is operated for the Department of Energy by Battelle.

References

1. DOE Hydrogen Program, www.hydrogen.energy.gov.
2. A. Gutowska, L. Li, Y. Shin, C. Wang, X. Li, J. Linehan, R.Smith, B. Kay, B. Schmid, W. Shaw, M. Gotowski and T. Autrey, *Angew. Chem., Int. Ed.*, **44**, (2005).
3. S. Rassat, R. Smith, T. Autrey, C. Aardahl, A. Chin, J. Magee, G. VanSciver and F. Lipiecki, *Prepr.Pap.-Am. Chem. Soc., Div. Fuel Chem.*, **51:517-9**, (2006).
4. C. Aardahl and S. Rassat, *International Journal of Hydrogen Energy*, **34(16)**, pp 6676-6683 (2009).
5. S. Rassat, C. Aardahl, T. Autrey and R. Smith, *Energy Fuels*, **24**, pp 2596-2606 (2010).
6. G. Lucas and W. Richards, *International Journal of Hydrogen Energy*, **9**(**3**), pp 225-231 (1984).
7. B. Hardy and D. Anton, *International Journal of Hydrogen Energy*, **34(5),** pp 2269-2277 (2009).
8. B. Hardy and D. Anton, *International Journal of Hydrogen Energy*, **34(7),** pp 2992-3004 (2009).
9. K. Brooks, M. Devarakonda, S. Rassat, D. King and D. Herling, *Proceedings of ASME 8th International Fuel Cell Science, Engineering and Technology Conference*, Brooklyn, NY (2010), *ISBN 978-0-7918-3875-4*.
10. Matlab and Simulink, www.mathworks.com (Last Accessed June 25th 2010).
11. Kristin Day (National Renewable Energy Laboratory), *Technical Communication*, (2010).
12. M. Devarakonda, J. Holladay, K. Brooks, S. Rassat, D. Herling, "Dynamic Modeling and Simulation Based Analysis of an Ammonia Borane Bead Reactor System for Hydrogen Storage", Vol 33, *"Polymer Electrolyte Fuel Cells 10", ECS Transactions, 2010.*

Hydrogen Generated From Hydrolysis of Ammonia Borane Using Cobalt and Ruthenium Based Catalysts

Cheng-Hong Liu[1], Yi-Chun Wu[1], Bing-Hung Chen[1,*], Chan-Li Hsueh[2], Jie-Ren Ku[2], Fanghei Tsau[2]

[1] *Department of Chemical Engineering, National Cheng Kung University, Tainan, 70101, TAIWAN*

[2] *New Energy Technology Division, Green Energy and Environment Research Laboratories, Industrial Technology Research Institute, Hsinchu, 31040, TAIWAN*

Stable and continuous supply of ultrapure hydrogen through controllable hydrolysis reaction of chemical hydrides, one kind of very promising hydrogen storage materials, has become more and more attractive recently, which has been shown as a promising and viable way to establishment of the hydrogen economy. Such chemical storage of hydrogen could give a much higher storage capacity of hydrogen. For example, ultrapure hydrogen can be released through hydrolysis of ammonia borane (NH_3BH_3, AB). Among various chemicals used as hydrogen storage media, chemical hydrides such as ammonia borane have attracted the most attention lately, especially after the US DOE concluded that sodium borohydride could reach the set 2010 goal on hydrogen storage capacity at 6 wt% but fails to achieve that set for 2015 at 9 wt%. The hydrogen storage capacity of ammonia borane is generally accepted as being able to reach above 9 wt% with its inherent hydrogen storage capacity as high as 19.6 wt%. More importantly, ammonia borane and its spent products after hydrogen production via hydrolysis reaction are rarely toxic, stable and easily handled at ambient conditions. However, to feasibly realize the hydrogen economics associated with the usage of ammonia borane as hydrogen storage materials, it is necessary to bring down the cost of ammonia borane. Consequently, in this proposal, we aim to identify the possible products of spent ammonia borane and to assess possible regeneration technology on the spent ammonia borane for a further use as media in hydrogen storage in order to cut the operation cost of such chemical storage system of hydrogen.

In this project, we aim to establish the way to characterize the product before and after the reaction of ammonia borane by hydrolysis and thermal decomposition for hydrogen production. Various equipments, such as FT-IR XRD NMR, were employed in this work. Furthermore, we have also attempted to use the NMR as well as the hydrogen release profiles from ammonia borane to determine and calibrate the portion of ammonia borane involved for hydrogen evolution, and, hence, to construct the possible regeneration path of spent ammonia borane back to AB.

With TGA results of synthesized Co/IR-120, Ru/IR-120 catalysts and pristine IR-120 resin beads, notably, metal-loaded catalysts show higher tolerance to temperature by *circa* 100°C, compared with resin beads, of which thermolysis temperature happens at about 300°C. Furthermore, it is also observed that amounts of cobalt and ruthenium loaded on the surface of supports, IR-120, are around 39 and 35 wt%, respectively. Effects of different initial loadings of Co/IR-120 catalysts

* *E-mail: bhchen@alumni.rice.edu*

on kinetics behaviors of ammonia borane (AB) hydrolysis were conducted as well. As results, the first order reaction model is proposed, according to the linear behavior found in $\ln([AB]/[AB]_0)$ *vs.* time. In addition, the product obtained from hydrolysis reaction of AB is also characterized to validate the relation between AB and $NaBH_4$. As found, the major product after hydrolysis is boric acid, which is beneficial for setting up a recycle mechanism between AB and $NaBH_4$.

An NMR Study of Sodium Borohydride Hydrolysis in Concentrated Aqueous Solution

AIChE Meeting November 2010

Lin Yu[1], Amy M. Beaird[1], Ping Li[2], J. Karl Johnson[2], Rajasree Retnamma[3], Michael A. Matthews[1]

1. Department of Chemical Engineering, University of South Carolina

2. Department of Chemical and Petroleum Engineering, University of Pittsburgh

3. National Laboratory for Energy and Geology, Lisbon, Portugal

1.Introduction

Sodium borohydride is the most widely studied material among chemical hydrides and has been identified as a promising hydrogen storage media for portable applications, because of advantages such as high theoretical energy density (10.8wt%), low cost, safe handling and ease of H_2 release. Traditionally, studies focus on dilute sodium borohydride solutions with the aid of catalysts at room temperature. Various kinetic models for catalyzed hydrolysis have been reported accordingly. However, no kinetic data are available for non-catalyzed, concentrated $NaBH_4$ solutions. This work reports an experimental kinetic study of non-catalyzed sodium borohydride hydrolysis in highly concentrated solutions at elevated temperatures. The concentrations of boron containing species were characterized in-situ by a novel ^{11}B NMR technique. The effects of temperature and pH are reported.

2. Experimental

A Varian Inova 500 spectrometer operating at 160.487Hz for ^{11}B was used to acquire spectra in real time. The NMR operates in the temperature range of 25-90°C, with an accuracy of +/-0.1 °C. ^{11}B NMR spectra were acquired as proton-coupled, with 15s relaxation time and 8 scans. The NMR spectrometer was tuned to increase the signal-to-noise ratio each time before each measurement was carried out.

A concentric cylinder arrangement was used for obtaining real-time measurements of the rate of appearance of $NaB(OH)_4$. Sodium borohydride powder was placed in an 8 mm NMR tube, which was then sealed in a glove box. The 8mm tube was placed inside a 10mm NMR tube containing a standard solution of boric acid dissolved in D_2O. Prior to beginning the NMR scan, a predetermined volume of DI water was injected to the 8mm tube to dissolve the $NaBH_4$. Hydrogen bubbles were immediately seen after DI water was introduced. The concentric tubes were immediately placed in the spectrometer. The elapsed time between solution addition and first NMR spectrum acquisition was in the range of 180 to 240s.

The peak intensities of the reactant $NaBH_4$ and product $NaB(OH)_4$ were collected as a function of time. The concentrations of $NaBH_4$ and $NaB(OH)_4$ were quantified with reference to the standard boric acid solution, utilizing the fact that the concentration is proportional to the ratio of tube cross-sectional areas and the ratio of peak intensities. The data were obtained at temperatures of 25°C, 60°C and 80°C and at $NaBH_4$ concentrations of 10wt%, 15wt% and 20wt%. Additional data were obtained for the 20wt% $NaBH_4$ solution at 80°C, using NaOH at concentrations from initial 1wt% to 15wt% to adjust the pH. Another set of data were obtained for 10wt% $NaBH_4$ at 25°C, using acetic acid and HCl to adjust the pH in the range of 2.85-4.69.

3. Representative Results and Discussion

The effect of initial $NaBH_4$ concentration on the rate of formation of the hydrolysis is shown in Figure1.The data presented are concentration of the product $NaB(OH)_4$ (the formula equivalent to $NaBO_2 \cdot 2H_2O$ in excess water) at 25°C; It is shown that increasing the initial $NaBH_4$ concentration (10 to 20wt%) gives increasing product concentration. Comparison of the product concentration after 24 hours to that of 100% conversion is indicated by horizontal lines. As shown in Figure 1, the most concentrated solution results in lower conversion. In order to quantify the effects of initial $NaBH_4$ concentration on reaction rate at 25 °C, we took the derivative of the product concentration with respect to time. We find that the reaction rate drops quickly within first 5 hours; and it increases with the increasing initial $NaBH_4$ concentration from 10wt% to 20wt%.

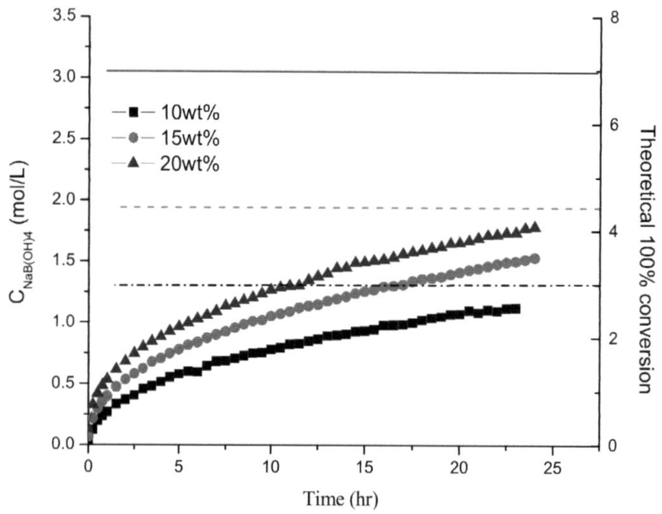

Figure 1. Effect of initial $NaBH_4$ concentration on formation of $NaB(OH)_4$ at 25°C

The effect of temperature within the range 25°C-80°C on the hydrolysis reaction has also been studied with an initial $NaBH_4$ concentration of 15wt%. Data will be presented during the presentation. It is found that at 80°C the reaction rate is the greatest; at around 10 hours, the $NaB(OH)_4$ has already reached theoretical value of 4.89mol/L for 100% conversion. However, at 25°C $NaB(OH)_4$ concentration only is 1.53 mol/L after 24 hours. Therefore, high temperature significantly improves the generation of product in the $NaBH_4$ hydrolysis. The reaction rate was observed to drop quickly within the first five hours and decrease gradually as the reaction continued. The initial reaction rate increases from 0.58 mol/L·hr to 6.73 mol/L·hr as temperature is increased from 25°C to 80°C.

To investigate the effect of increasing pH on the kinetics, alkaline solutions with different initial NaOH concentrations (1wt%, 5wt%, 10wt% and 15wt%) were prepared, with a constant initial $NaBH_4$ concentration of 20wt% at 80°C. NaOH inhibits the hydrolysis reaction. The inhibition effect is the least evident at 1wt% of NaOH, but this effect becomes more pronounced by increasing the concentration of NaOH from 5wt% to 15wt%. For the initial 15wt% NaOH, the hydrolysis rate is constant, very near zero.

The acceleration of the reaction with acid was accomplished by adjusting the initial pH. We used acetic acid and HCl to adjust the initial pH over the range from 2.85 to 4.69 to investigate the hydrolysis kinetics. It is found that with the same initial pH, acetic acid is a better accelerator than HCl at the same conditions.

To fit the experimental kinetic data, we fit the empirical power law model shown in equation (1). We assumed that both the $NaBH_4$ concentration and proton concentration will affect the hydrolysis rate:

$$R= \frac{dCNaB(OH)_4}{dt} = [NaBH_4]^m[H^+]^n \quad (1)$$

The kinetic parameters were obtained by a linear regression of the rate versus $NaBH_4$ concentration. The rate equation is based on the assumption of a constant water concentration during the hydrolysis reaction. We are able to correlate the measured pH with the product concentration under the same reaction conditions. For example, Table 1 shows the regression results from the left division algorithm in Matlab. The reaction order with respect to $NaBH_4$ decreases with increasing temperature from 25°C to 60°C; while the order with respect to proton concentration remains unaffected to temperature change in this range. The rate constant also shows an increase with rising temperature.

Table 1 Kinetic Parameters Determined by Linear Regression (neutral pH)

T	m	n	logk
25°C	2.8	1.1	7.5
60°C	1.9	1.1	9.5

We also performed a linear regression for the reaction order with respect to $NaBH_4$ and NaOH at 80°C, respectively. According to the result, the reaction rate model can be expressed by model equation (2):

$$R = k\,[NaBH_4]^{3.3}[NaOH]^{-0.92} \quad (2)$$

It is seen that the reaction rate increases with the $NaBH_4$ concentration and decreases with NaOH concentration. The negative dependence of reaction rate on the NaOH concentration is consistent with the inhibition effect of NaOH.

4. Conclusion

In this study, we investigated the non-catalyzed hydrolysis kinetics of concentrated $NaBH_4$ and NaOH stabilized solutions for hydrogen generation. The reactions are evaluated using ^{11}B NMR at 25, 60 and 80°C for 24 hrs. It is found that both high initial $NaBH_4$ concentration and high temperature are able to increase the reaction rate of $NaBH_4$ aqueous hydrolysis. Concentrated alkaline solutions will result in severe inhibition of the hydrolysis reaction. In contrast, acids accelerate the hydrolysis rate. We have also developed an empirical kinetic power-law model for the hydrolysis. It indicates that at higher temperature the dependence of hydrolysis reaction rate on $NaBH_4$ concentration decreases; the increasing NaOH concentration decreases the reaction rate. Future work will take into account the water concentration change in modeling.

(336a) Achieving Optimal Hydrogen Storage In MOF-5

Andrea Sudik[1], Jun Yang[2], Justin Purewal[3], Donald J. Siegel[3], Emi Leung[4] and Ulrich Mueller[5], (1) Fuel Cell and Hybrid Electric Vehicle Research, Ford Motor Company, Dearborn, MI, (2)Fuel Cell and Hybrid Electric Vehicle Department, Ford Motor Company, Dearborn, MI, (3)Mechanical Engineering Department, University of Michigan (Ann Arbor), Ann Arbor, MI, (4)Chemicals Research and Engineering, BASF SE, Ludwigshafen, Germany, (5)Chemicals Research and Engineering, Ford Motor Company, Dearborn, MI

Storage of hydrogen in condensed phases has the potential to be more efficient and less expensive than the most widely used storage method, compressed gaseous hydrogen. While substantial progress has recently been made in identifying improved hydrogen storage materials, significant challenges associated with the engineering of the storage system around a candidate storage material persist. In particular, the degree to which materials processing can act as a countermeasure to improve upon the deficiencies of a given storage material is largely unknown. This paper will detail recent efforts aimed at determining the processing-structure-properties (PSP) relationships for an important class of sorbent materials [metal organic frameworks (MOFs)] in order to devise improved packing and processing strategies for their use in a hydrogen storage systems context. In particular, we will detail various mechanical processing routes that have been explored (ranging from powders to pelletization) in an effort to simultaneously maximize packing density, heat and mass transfer, and hydrogen uptake characteristics. We will specifically highlight compaction studies for Basolite Z100H (the commercial form of MOF-5) wherein the relevant hydrogen storage properties have been characterized as a function of processing route for the identification of preliminary PSP relationships. The intent of this investigation is to identify rational pathways by which materials properties can be optimized towards achieving commercially-viable hydrogen storage systems, and to reveal the upper limits of performance achievable with current-generation hydrogen storage materials.

See more of this Session: Hydrogen Storage System Engineering and Applications: Applied Materials Development III
See more of this Group/Topical: Topical 8: Hydrogen Production and Storage

CONFERENCE ABSTRACT

Improving Comparability of Hydrogen Storage Capacities of Nanoporous Materials

V. P. Ting, N. Bimbo, A. Hruzewicz-Kołodziejczyk, L. Fisher, S. Rigby,
A. Burrows and T. J. Mays

One of the primary barriers to the successful and timely implementation of a "Hydrogen Economy" with hydrogen as an energy carrier is the ability to efficiently store hydrogen for later use. The storage of hydrogen via physisorption of H_2 onto porous solid state materials benefits from reversible and rapid adsorption and desorption at moderate temperatures and pressures compared to either chemical storage or storage of hydrogen as a compressed gas or cryogenic fluid. The search for the optimal porous solid state storage medium exhibiting acceptably high volumetric and gravimetric densities has long been the focus of intense research. Due to a lack of standardized data collection and analysis routines, it is difficult, if not impossible, to directly compare the hydrogen storage capacities of such materials, as the real material properties cannot be decoupled from the influence of the different data collection or modeling methods used. We present our investigations into producing a standardized routine for the assessment of the hydrogen storage capacities of different nanoporous materials. This includes the formulation of rigorous and systematic data collection routines for the purpose of obtaining universally reproducible H_2 isotherms under sub-atmospheric, and high pressure conditions, as well as new methods for modelling the resultant isotherms in order to extract information on the interactions of H_2 gas with the porous solid and gain an improved understanding of the physisorption process itself. We will demonstrate these practices, as applied to the analysis of some well-characterized standard materials including commercially-available zeolites and tailored activated carbons materials. With this greater knowledge of the variables that affect the quality of hydrogen sorption data, the resultant routines will permit better control of the myriad contributing experimental factors. Combined with the improved data modelling methods, it will become possible to directly compare the hydrogen storage capabilities of any nanoporous materials, a development which will prove invaluable in the evaluation of hydrogen storage materials investigated in the future.

(336c) Thermal Stability of Nanopores in Palladium Alloys and Their Hydrides

David B. Robinson[1], Markus D. Ong[1], Benjamin W. Jacobs[1], Mary E. Langham[1], Michael S. Kent[2] and Ilke Arslan[3], (1)Energy Nanomaterials, Sandia National Laboratories, Livermore, CA, (2)Sandia National Laboratories, Albuquerque, NM, (3)Dept. of Chemical Engineering and Materials Science, University of California, Davis, Davis, CA

Nanoporous material architectures can offer improved charging and discharging kinetics in metal hydrides due to their high surface area. We have synthesized nanoporous palladium and palladium alloy powders in a scalable fashion by reduction of palladium salts in a concentrated aqueous surfactant. Particle diameters are micrometer-scale, and each particle is perforated by 3 nm pores. In pure palladium, the pores are unstable at 150 C, but when alloyed with rhodium, they are stable up to 400 C on short timescales. These results are expected to be related to pore lifetime at more moderate temperatures. Electron tomography and elemental mapping after in situ or ex situ heating and hydrogen exposure help us understand the mechanism of the improvement. We observe surface segregation of the minor component in alloys. Pore density and stability are also related to the microstructure of the material. Improved stability can be obtained with minimal sacrifice of hydrogen storage properties.

See more of this Session: Hydrogen Storage System Engineering and Applications: Applied Materials Development III
See more of this Group/Topical: Topical 8: Hydrogen Production and Storage

(336d) Facilitating Dehydrogenation of Complex Hydride Hydrogen Storage Materials Using a CNTs-Based Catalytic Matrix

Sean S.-Y. Lin[1], Jun Yang[2], Xin Zhao[1] and Harold H. Kung[1], (1)Chemical and Biological Engineering, Northwestern University, Evanston, IL, (2)Research and Advanced Engineering, Ford Motor Company, Dearborn, MI

Complex hydrides are considered one of the most promising hydrogen storage materials. Hydrogen evolution from complex hydrides is dictated by an inter-related sequence of chemical reaction steps including decomposition of the hydride, diffusion of hydrogen atom/ion and counter ions in the solid lattice, surface diffusion of these atoms/ions, recombination of hydrogen atoms, transformation of lattice phases, and desorption of hydrogen molecules. Using carbon nanomaterials such as carbon fibers, carbon nanotubes, and carbon fullerenes as additive has shown promising result to facilitate dehydrogenation of complex hydrides. Several possible mechanisms have been proposed to explain the effect, including size reduction of hydride particles, confinement effect that prevents phase segregation, and electronic effect that destabilizes the covalent bond associated with hydrogen in the hydrides.

In this study, a light weight, high surface area CNTs-based catalytic matrix was developed and investigated for hydride decomposition using two model complex hydrides, NaAlH4 and Ca(BH4)2. Adding a non-precious metal (Co, Ni, Fe) in the form of nanoparticles decorating the carbon matrix (as confirmed by SEM images) to the complex hydrides facilitated hydride decomposition. The decomposition temperature of NaAlH4 was decreased from 260 oC to 210 oC by adding 10 wt% CNTs only. The temperature was further decreased to 180 oC with a Co-decorated CNTs matrix. In the case of Ca(BH4)2, the decomposition temperature was decreased from 376 to 366 oC. The CNTs-based catalytic matrix was more effective in NaAlH4 decomposition possibly because a better hydride-catalyst interface was created between the melted NaAlH4 and the catalytic matrix by capillary action. Such interface benefits the overall kinetics by facilitating the chemical bond breakage as well as mass transfer of the product nuclei, which can be dispersed homogeneously via the liquid phase. By dispersing such a composite catalyst in a mixture with complex hydride, we can attain a high contact area between the catalyst and the hydride with minimal weight penalty. The contact points could serve both for hydrogen transfer between the catalyst complex and the hydride and as potential nucleation points for phase transformation of the hydride.

See more of this Session: Hydrogen Storage System Engineering and Applications: Applied Materials Development III
See more of this Group/Topical: Topical 8: Hydrogen Production and Storage

Carbon Nanostructures as Catalyst for Improving the Hydrogen Storage Behaviour of Complex Aluminium Hydride

*M. Sterlin Leo Hudson, Himanshu Raghubanshi, D Pukazhselvan and O. N. Srivastava**
Nanoscience and Technology Unit, Department of Physics, Banaras Hindu University, Varanasi, India
* **Corresponding author** : Email: hepons@yahoo.com; Tel: 0542-2368468; Fax: 0542-2369889

Abstract:

The present paper reports the catalytic effect of carbon nanostructures, particularly graphitic nanofibres (GNFs) with different structure morphology, namely helical GNFs (HGNFs) and planar GNFs (PGNFs) as the catalyst for complex aluminium hydride. HGNFs and PGNFs were synthesized by catalytic thermal decomposition of acetylene (C_2H_2). The growth of HGNFs was achieved by employing faceted Ni nanoparticles, whereas spherical Ni nanoparticles produce PGNFs. It has been observed that HGNFs posses superior catalytic activity than PGNFs in improving the desorption kinetics and decreasing the desorption temperature of complex aluminium hydride ($NaAlH_4$, $LiAlH_4$ and $LiMg(AlH_4)_3$). Temperature programmed desorption (TPD) reveals that HGNFs admixed alanates undergo hydrogen desorption at a much lower temperature than PGNFs admixed material. Thus for the heating rate of 2°C/min, the peak desorption temperature corresponds to initial step desorption of $NaAlH_4$ admixed with 8 mol% HGNFs and 8 mol% PGNFs has been lowered to 143.6°C and 152.6°C, respectively. In addition to the enhancement in desorption kinetics, the GNFs admixed $NaAlH_4$ also undergoes rehydrogenation at the moderate condition. In order to get supportive evidence for our experiment, we have carried out ab-initio studies by calculating hydrogen removal energy from alanates with planar and helical model of GNF. It becomes clear that from both experiment and ab-initio calculations, that catalytic effect of GNFs is curvature dependent. Thus HGNF with higher helicity helps to lower the energy needed to remove hydrogen from alanates.

Keywords: Complex hydrides, carbon nanostructures, catalysts, temperature programmed desorption

1. Introduction:

Complex aluminium hydrides, $NaAlH_4$, LiAlH4 and $LiMg(AlH_4)_3$ possess the quite high gravimetric hydrogen capacity of 7.3, 10.6 and 9.7 wt.%H_2, respectively. The fast desorption of hydrogen from these compounds becomes feasible with the use of suitable catalysts. The growing interest on finding an effective catalyst for improving the dehydrogenation/re-hydrogenation behavior of complex metal hydrides after the pioneering work of Bogdanovic et al [1] on the reversibility of $NaAlH_4$ triggers research on various catalysts, including transition and rare earth metal based catalysts. There is an ongoing search for finding an effective catalyst for improving the dehydrogenation/re-hydrogenation behaviour of complex hydrides [2]. Recent studies by Berseth et al [3] reveal that carbon nano-variants are found to possess catalytic effect on complex hydrides. Among the different carbon nanovariants, graphitic nanofibres are very promising due to their large number of localized σ bond electrons around the edges of parallel graphitic sheets, which provides more catalytically active sites [4, 5]. Advantageous features of carbon nanofibre are that it is light weight, not prone to oxidation nor gets substituted in the alanate lattice. In the light of above, we explored the use of graphitic nanofibres of two different morphology namely planar graphitic nanofibre (PGNF) and helical graphitic nanofibre (HGNF) for possible improvement in dehydrogenation characteristics of $NaAlH_4$, $LiMg(AlH_4)_3$ and

$LiAlH_4$. The catalytic activity of GNFs has been proven for binary hydride, MgH_2 [6]. The optimum concentration of GNFs on $NaAlH_4$, $LiAlH_4$ and $LiMg(AlH_4)_3$ for the present study is 8 mol%. The catalytic effect of GNFs on $NaAlH_4$ has been compared with single walled CNTs (SWCNTs). Our investigation reveals that HGNFs possess superior catalytic activity than PGNFs and SWCNTs.

2. Experimental:

As received $NaAlH_4$ (Aldrich, 93%), $LiAlH_4$ (Aldrich, 95%) and $MgCl_2$ (Alfa Aesar, 99%) have been used without further purification. $LiMg(AlH_4)_3$ was synthesized by ball-milling $LiAlH_4$ and $MgCl_2$ in the molar ratio 3:1 together with two steel balls of 12.5 mm each and one ball with 4 mm in a chrome-nickel stainless steel milling vial under argon atmosphere using a locally fabricated attritor ball-miller. The ball-to-powder mass ratio was kept at 10:1. The expected reaction $3LiAlH_4 + MgCl_2 \rightarrow LiMg(AlH_4)_3 + 2LiCl$ has been found to get completed in one hour of ball-milling. All operations on the samples were done under dry argon atmosphere in a glove box to prevent reaction with moisture and oxygen in the air. The graphitic nanofibres are being synthesized in our laboratory through dissociation of hydrocarbon (acetylene/ethylene) over Ni/Fe nanoparticles catalyst at 650°C under flowing He and H_2 gas both with an optimized flow rate of 1500 sccm. In keeping with the known results [7, 8], we have found that morphology of GNF is dependent on the shape of the Ni/Fe nanoparticles. For faceted Ni/Fe nanoparticles helical GNF get formed and for non-faceted Ni/Fe nanoparticles, planar GNF is found to grow. SWCNTs were supplied by MK IMPEX CANADA. Hydrogen desorption behaviour of $NaAlH_4$, $LiAlH_4$ and $LiMg(AlH_4)_3$ samples were measured using computerized pressure composition isotherm (P-C-I) measurement system supplied by Advanced Materials Corporation.

Structural characterization of the samples was carried out through XRD, Philips PW-1710 and X'Pert PRO (PAN*alytical*) X-ray diffractometer equipped with the graphite monochromator employing CuK_α radiation (λ= 1.5402Å). Exposure of the sample to the atmosphere has been avoided by covering the XRD holder by fine layer of Parafilm (Pechiney plastic packing). The microstructural characterization has been done through Transmission electron Microscope (Technai, 200 kV).

3. Results and Discussions:

3.1. Sodium Aluminium Hydride:

$NaAlH_4$ desorbs hydrogen through the following reactions,

$3NaAlH_4 \rightarrow Na_3AlH_6 + 2Al + 3H_2$ (170°C) (1)
$Na_3AlH_6 + 2Al \rightarrow 3NaH + 3Al + 3/2H_2$ (250°C) (2)
$3NaH + 3Al \rightarrow 3Na + 3Al + 3/2H_2$ (>400°C) (3)

Reaction (1) and (2) releases 3.7 and 1.8 wt.% H_2, respectively. Both reaction (1) and (2) liberate 5.5 wt.% H_2, satisfying the US-DOE revised target of 5.5 wt.% H_2 [9]. However, an effective catalyst is required to make reaction (1) and (2) reversible under moderate temperature and pressure. Reaction (3) is generally not considered usable for hydrogen storage, due to its very high temperature. We have employed carbon nanovariants of different types, namely HGNFs, PGNFs and SWCNTs as the catalyst for improving the dehydrogenation and rehydrogenation behaviour of $NaAlH_4$. Representative TEM micrographs of HGNF and PGNF are shown in Figs. 1(a) and 1(b).

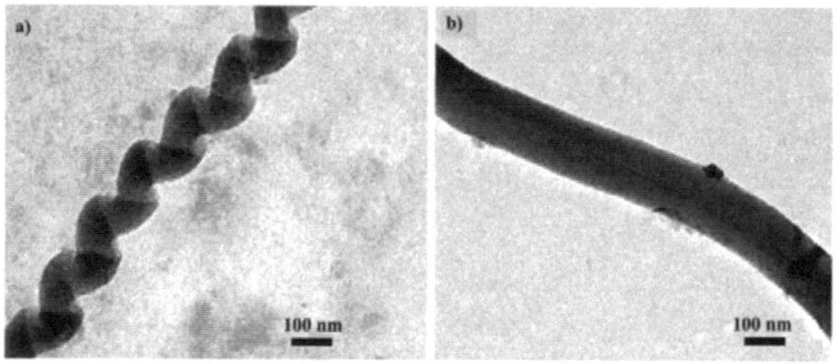

Fig. 1. TEM microgram of GNF a) helical morphology GNF (HGNF) and b) planar morphology GNF (PGNF) (difference between HGNFs and PGNFs are given in table 1)

Table 1: Difference between HGNF and PGNF used as catalyst

Types of GNFs	Catalyst used for synthesis	Surface area (m^2/g)	Average diameter (nm)
Helical GNF (HGNF)	Faceted Ni nanoparticle	125	150
Planar GNF (PGNF)	Spherical Ni nanoparticle	109	200

The temperature programmed desorption (TPD) curve of $NaAlH_4$ admixed with different carbon nanovariants is shown in Fig. 2. As it becomes evident from Fig. 2 that for the heating rate of 2°C/min, the peak desorption temperature of pristine $NaAlH_4$ corresponds to Eq. (1) and Eq. (2) are 163°C and 240°C. However, admixing of 8 mol% carbon nanovariants with $NaAlH_4$ lowers the desorption temperature of $NaAlH_4$. The peak desorption temperature of $NaAlH_4$ admixed with 8 mol% HGNF corresponds to Eq. (1) and Eq. (2) are 143.6°C and 190°C, respectively. Whereas that for $NaAlH_4$ admixed with 8 mol% PGNF corresponds to Eq. (1) and Eq. (2) is 152.6°C and 209°C, respectively. It has been observed that among the different forms of CNTs and GNFs, HGNFs posses superior catalytic effect in lowering the desorption temperature corresponds to both Eq (1) and Eq(2) of $NaAlH_4$. The dehydrogenation kinetics and rehydrogenation kinetics of $NaAlH_4$ admixed with 8 mol% HGNF is found to be superior to that of PGNF and CNT admixed sample.

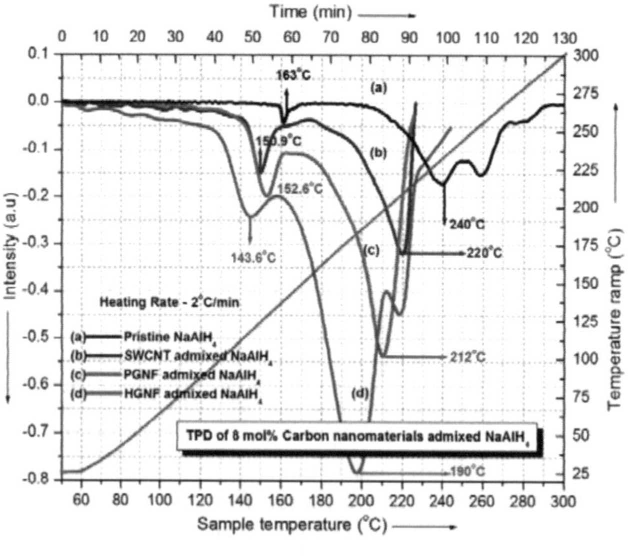

Fig. 2. TPD curve of 8 mol% carbon nanovariants admixed $NaAlH_4$ for the heating rate of 2°C/min (red line represents the heating profile).

3.2. Lithium-magnesium alanate:

$LiMg(AlH_4)_3$ has been synthesized mechano-chemically through the following reaction.

$3Li(AlH_4)_3 + MgCl_2 \rightarrow LiMg(AlH_4)_3 + 2LiCl$ (4)

by using a locally fabricated attritor ball-miller. The above reaction gets completed in ~1 hour of ball-milling. Following the known results [10], the thermal decomposition behavior of $LiMg(AlH_4)_3$ can be described as follows,

$LiMg(AlH_4)_3 \rightarrow LiMgAlH_6 + 2Al + 3H_2$ (~100°C) (5)

$LiMgAlH_6 \rightarrow LiH + MgH_2 + Al + 3/2H_2$ (~150°C) (6)

$2MgH_2 + 3Al \rightarrow Al_3Mg_2 + 2H_2$ (~340°C) (7)

The temperature programmed desorption (TPD) curve was monitored from room temperature to 240°C at the heating rate of 1°C/min. It has been observed from the TPD profile that the peak decomposition temperature of Eq. (5) is 105.5°C and that of Eq. (6) is 149.6°C. For 8 mol% HGNF admixed $LiMg(AlH_4)_3$, Eq. (5) occurs at the time of catalyst admixing. However, for the heating rate of 1°C/min, the peak desorption temperature for 8 mol% HGNF admixed $LiMg(AlH_4)_3$ corresponds to Eq. (6) has been lowered from 149.6°C to 127.9°C. A representative TPD curve is shown in Figs. 3.

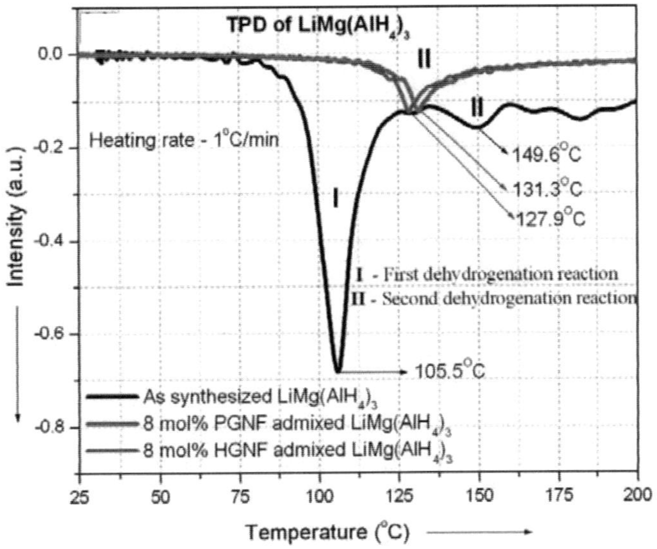

Fig. 3. Temperature Programmed Desorption (TPD) of $LiMg(AlH_4)_3$ for the heating rate of 1°C/min

In order to find the catalytic effect of HGNF in decreasing the activation energy of $LiMg(AlH_4)_3$, we proceed to evaluate this energy from the Arrhenius plot, as described by Janot et al. [11].

The activation energy for the desorption process is estimated by plotting the maximum reaction rate (*k*) as a function of the temperature (*T*). The reaction rates are obtained from the slopes of the tangents at the inflection points of desorption kinetic curves. The maximum reaction rates (*k*) of Eq. (5) at the different temperatures were calculated from desorption kinetic curves as shown in Fig. 4(a). Since the inclusion of LiCl mass in the kinetic measurement of $LiMg(AlH_4)_3$ affects the rate constant value, we have not considered the weight of LiCl for desorption kinetic measurements.

Quantitative estimation of kinetic barrier was then carried out by the determination of the activation energy (*Ea*) from the well known Arrhenius equation of the form

$$\ln(k) = -Ea/RT + \ln(A) \quad (8)$$

By plotting ln *k* versus *1/T*, *Ea* can be extracted from the linear slope. Fig. 4(b) shows the Arrhenius plot for the first-step dehydrogenation of $LiMg(AlH_4)_3$. A good linearity between ln *k* and 1/*T* is obtained with R^2 = 0.99. The activation energy (*Ea*) as calculated corresponds to ~66 kJ/mol.

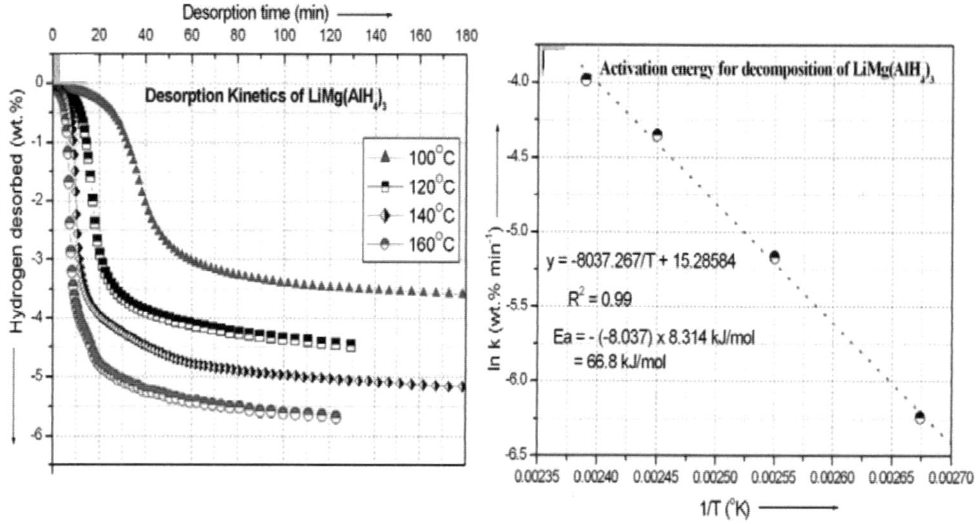

Fig. 4. a). desorption kinetic curve of $LiMg(AlH_4)_3$ at different temperatures b) plot of ln *k* vs 1/T for $LiMg(AlH_4)_3$.

3.3. Lithium alanate:

In general, $LiAlH_4$ has three-stage decomposition, as represented by the following equations:

$3LiAlH_4 \rightarrow Li_3AlH_6 + 2Al + 3H_2$ (~150°C) (9)

$Li_3AlH_6 \rightarrow 3LiH + Al + 3/2H_2$ (~180°C) (10)

$3LiH + 3Al \rightarrow 3LiAl + 3/2H_2$ (>400°C) (11)

Eq. (9) approximately releases 5.3 wt.% hydrogen and eq. (10) and eq. (11) both release approximately 2.6 wt.% H_2 each [12].

The catalyst admixing was carried out through ball-milling for 15 min, using the same ball-miller as mentioned above. The optimum molar concentration of GNFs for $LiAlH_4$ is found to be 8 mol%. Lower concentration of GNF does not produce enhancement in kinetics similar to that from 8 mol% and increasing the molar concentration of GNFs further leads to the decrease in hydrogen storage capacity. Fig. 5 shows the representative TPD profile of $LiAlH_4$ sample. There is a loss of ~2 wt.% H_2 in the 8 mol% GNFs admixed $LiAlH_4$ sample. This was estimated by comparing the TPD profile of the ball-milled and pristine $LiAlH_4$.

A noticeable feature shown in Fig. 5 is that for the heating rate of 1°C/min, the peak desorption temperature for 8 mol% HGNF admixed $LiAlH_4$ sample corresponds to Eq. (6) has been lowered from 158.8°C to 128.5°C and as that for reaction corresponding to Eq. (7) has been lowered from 206.1°C to 165.5°C. It may be mentioned that the effect of an increase in the interface area which will result due to ball-milling, will not be able to explain the significant decrease in desorption temperature. Apparently, it is the catalytic effect of GNF, which leads to decrease of desorption temperature. The exact role of GNF in improving the dehydrogenation behaviour of complex hydride is exactly not known. Based on the known results on the catalytic effect of carbon nanostructures in complex hydrides, it is believed that higher the curvature of catalyst, higher is the catalytic activity. The carbon nanostructures as for example fullerenes have higher curvature than CNTs or graphenes.

Hence fullerene has higher catalytic activity [3]. In contrast to PGNF, HGNF has curved surface. Therefore it will have higher catalytic activity than PGNF.

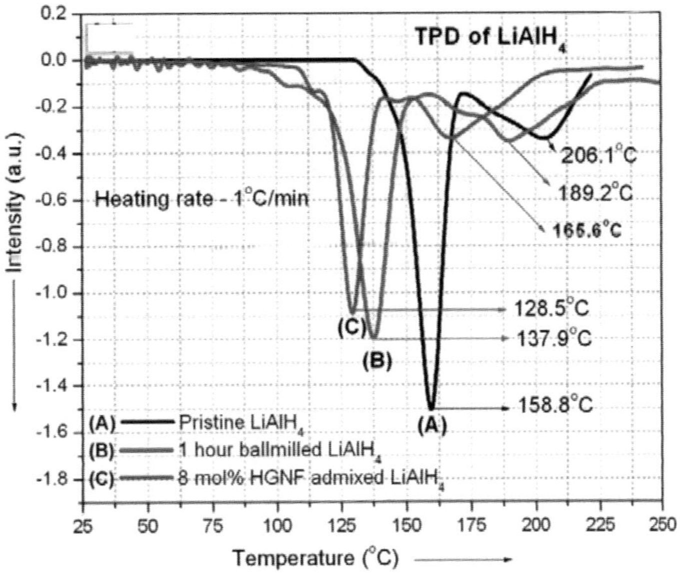

Fig. 5. Temperature programmed desorption of $LiAlH_4$

When admixing 8 mol% HGNF with $LiAlH_4$, desorption kinetics of $LiAlH_4$ get significantly improved. The effect of HGNFs on $LiAlH_4$ is estimated from the activation energy value (*Ea*) calculated for pristine $LiAlH_4$ and 8 mol% HGNF admixed with $LiAlH_4$. In order to find the activation energy of pristine $LiAlH_4$ using Arrhenius relation, the desorption kinetics of pristine $LiAlH_4$ at different temperatures were evaluated. The activation energy calculated for pristine $LiAlH_4$ is 107 kJ/mol. This value is comparable to the recently reported value of 102 kJ/mol by Blanchard et al.[13].

The activation energy (*Ea*) calculated for 8 mol% HGNF admixed $LiAlH_4$ is ~68 kJ/mol. Thus the admixing of 8 mol% HGNF helps to decrease the activation energy for the decomposition of $LiAlH_4$ by 36% (for pristine $LiAlH_4$ calculated *Ea* is 107 kJ/mol, and for 8 mol% HGNF admixed $LiAlH_4$, it is ~68 kJ/mol.

In order to get supportive evidence for our experiment, we have carried out ab-initio molecular simulation studies using density functional theory by calculating hydrogen removal energy from $NaAlH_4$ and $NaAlH_4$ with planar and helical model of GNF. This also confirms that planar model of GNF exhibit the superior catalytic effect due to its increased elecronegativity. The detailed results will be communicated elsewhere.

4. Conclusions:

The catalytic effect of graphitic nanofibres of two morphology, namely HGNF and PGNF on $NaAlH_4$, $LiMg(AlH_4)_3$ and $LiAlH_4$ has been studied. The dehydrogenation behaviour of the samples was studied through TPD and desorption kinetic measurements. We have found that helical morphology of GNF improves the dehydrogenation behaviour of both $LiMg(AlH_4)_3$ and $LiAlH_4$. For the heating rate of 2°C/min, the lowering of peak desorption temperature of $NaAlH_4$ admixed with 8 mol% HGNF as compared to its pristine phase is from 163°C to 143°C for the initial step and from 240°C to 190°C for the second step. Similarly, for $LiAlH_4$ and $LiMg(AlH_4)_3$, HGNF is found to be an effective catalyst for lowering the dehydrogenation temperature. The activation energy calculated for the first step decomposition of HGNF admixed $LiAlH_4$ is ~68 $kJmol^{-1}$, whereas that for pristine $LiAlH_4$, it is ~107 kJ/mol.

Acknowledgement:

The authors would like to thank Prof. T.N. Veziroglu, Prof. C.N.R. Rao, Prof. S.K. Joshi, Prof. S.P. Thyagarajan and Prof. D.P. Singh (VC:BHU) for their encouragement and support. Financial support from the Ministry of New and Renewable Energy, the University Grants Commission, DST: Nano Science Technology Unit (BHU) and the Council of Scientific and Industrial Research are thankfully acknowledged.

References:

1. Bogdanovic B, Schwickardi M. Ti-doped alkali metal aluminium hydrides as potential novel reversible hydrogen storage materials. J. Alloys Compd. 1997;253-254:1-9.
2. Srinivasan S, Escobar D, Goswami Y, Stefanakos E. Effects of catalysts doping on the thermal decomposition behavior of $Zn(BH_4)_2$. Int. J Hydrogen Energy. 2008;33:2268 -72.
3. Berseth PA, Harter AG, Zidan R, Blomqvist A, Araujo CM, Scheicher RH, Ahuja R, Jena P. Carbon Nanomaterials as Catalysts for Hydrogen Uptake and Release in $NaAlH_4$. Nano Lett. 2009;9:1501-05.
4. Serp P, Corrias M, Kalck P. Carbon nanotubes and nanofibers in catalysis. Applied Catalysis A: General. 2003;253:337-58.
5. Kumar LH, Viswanathan B, Murthy SS. Dehydriding behaviour of $LiAlH_4$-the catalytic role of carbon nanofibres. Int. J Hydrogen Energy. 2008;33:366-73.
6. Lillo-Rodenas MA, Guo ZX, Aguey-Zinsou KF, Cazorla-Amoros D, Linares-Solano A. Effects of different carbon materials on MgH_2 decomposition. Carbon 2008;46:126-37.
7. Park C, Baker RTK. Catalytic Behavior of Graphite Nanofiber Supported Nickel Particles. 2. The Influence of the Nanofiber Structure. J. Phys. Chem. B 1998;102:5168-77.
8. Blank VD, Alshevskiy YL, Zaitsev AI, Kazennov NV, Perezhogin IA, Kulnitskiy BA. Structure and phase composition of a catalyst for carbon nanofiber formation. Scripta Mat. 2006;55:1035-38.
9. Yang J, Sudik A, Wolverton C and Siegel D.J. High capacity hydrogen storage materials: attributes for automotive applications and techniques for materials discovery. Chem Soc Rev. 2010;39:656-675.
10. Tang X, Opalka SM, Laube BL, Wu FJ, Strickler JR, Anton DL. Hydrogen storage properties of Na–Li–Mg–Al–H complex hydrides. J. Alloys Compd. 2007;446-447:228-31.
11. Janot R, Eymery JB, Tarascon JM. Investigation of the processes for reversible hydrogen storage in the Li–Mg–N–H system. Journal of Power Sources. 2007;164:496-502.
12. Vittetoe AW, Niemann MU, Srinivasan SS, McGrath K, Kumar A, Goswami DY, Stefanakos EK, Thomas S. Destabilization of $LiAlH_4$ by nanocrystalline MgH_2. Int. J Hydrogen Energy. 2009;34:2333-39.
13. Blanchard D, Brinks HW, Hauback BC, Norby P, Muller J. Isothermal decomposition of $LiAlD_4$ with and without additives. J. Alloys Compd. 2005;404-406:743-47.

    Chen J, Kuriyama N, Xu Q, Takeshita HT, Sakai T. Reversible Hydrogen Storage via Titanium-Catalyzed $LiAlH_4$ and Li_3AlH_6. J. Phys. Chem. B. 2001;105:11214

Hydrogen Storage and Effects of Cycling on Li-Based Hydrides

Wen-Ming Chien*, Joshua H. Lamb and Dhanesh Chandra

Chemical and Materials Engineering / MS388, University of Nevada, Reno, Reno, NV 89557

Abstract

The hydrogen storage capacity was evaluated on lithium imide/amide hydriding/de-hydriding pressure cycling by using hydrogen with low level impurity gases (100 ppm O_2, H_2O, CH_4 or N_2) and industrial hydrogen (H_2O~32ppm, O_2~10ppm and others). Pressure-composition isotherms were obtained at 255°C after cycling between $Li_2NH \leftrightarrow LiNH_2$. Recent research found that Li-based hydrides are candidates for reversible hydrogen storage [2-5], and lithium nitride (Li_3N) has a theoretical capacity of up to ~11 wt% hydrogen when fully converted to $LiNH_2$. Chen et al. [6-7] have shown that Li_3N will absorb up to 11.5 wt% hydrogen with a reversible capacity of 7 wt%. Other research has shown the improvement of the reversible hydrogen storage properties by mixing different ratios of $LiNH_2$ / LiH [8] and $LiNH_2$ / Li_3N [9]. Huq et al. [10-11] have performed structural studies and in-situ deuteration/dedeuteration structural studies of Li_3N using powder neutron diffraction. Chien et al. [12] reported the phase transformation behaviors of Li_2ND, LiD and $LiND_2$ phases during the hydriding/dehydriding of Li_3N.

In this study, the losses of hydrogen storage capacities after pressure cycling were determined. The pressure-composition isotherm results shown that after 560 pressure cycles, 100 ppm O_2 in hydrogen were most detrimental (0.4 wt%H_2 capacity remaining out of ~5.6 wt.%) as compared to 100 ppm H_2O in hydrogen gas (2 wt%H_2). It was found that 1.65 wt.% hydrogen remained after 500 cycles by using industrial hydrogen. Pressure isotherms after ~100 pressure cycles showed 3.6 and 3.3 wt%H_2 capacity remained by using 100 ppm CH_4 and industrial hydrogen, respectively. X-ray diffraction results of the products after pressure cycling showed mainly Li_2NH and LiH phases, and the impurity Li_2O phase. The phase analysis results showed the Li_2NH phase reduced from 77% to 13%, and LiH phase increases from 18% to 57% after 1100 cycles. A summary plot pressure-composition isotherms (at 255°C) before and after Li_2NH-$LiNH_2$ cycling (no catalyst) with industrial hydrogen and 100 ppm levels of O_2, H_2O, CH_4 in UHP H_2 shows in Figure 1. The results showed that the nitrogen addition both allowed more of the theoretical hydrogen capacity to be storage as well as improved the stability of the hydrides after up to 1359 cycles. X-ray diffraction data shows that more Li_2NH phase (less LiH phase) is present in the desorbed hydride after cycling with the nitrogen addition than is present without. The results of the effect of the impurity gases in hydrogen showed that the N_2 addition will improve the hydrogen storage capacity, and the O_2 has the most detrimental effect.

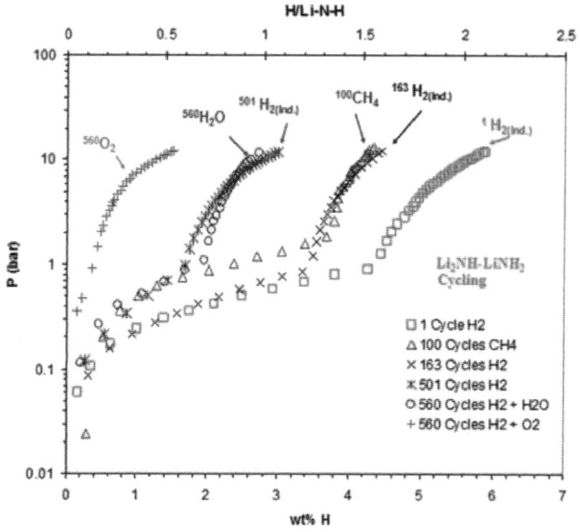

Figure 1. Summary plot of pressure-composition isotherms (at 255°C) before and after Li_2NH-$LiNH_2$ cycling (no catalyst) with industrial hydrogen and 100 ppm levels of O_2 , H_2O, CH_4 in UHP H_2.

References:

[2] Y. Nakamori and S. Orimo, *Materials Science and Engineering*, B108, 2004, p 48-50.

[3] Y. Nakamori and S. Orimo, *J. Alloys Compd.*, 370, 2004, p 271-275.

[4] T. Ichikawa, S. Isobe, N. Hanada and H. Fujii, *J. Alloys Compd.*, 365, 2004, p 271-276.

[5] J.W. Wiench, V.P. Balema, V.K. Pecharsky and M. Pruski, *J. Solid State Chemistry*, 117, 2004, p 648-653.

[6] P. Chen, Z. Xiong, J. Luo, J. Lin and K.-L. Tan, *Nature*, 420, 2002, p 302-304.

[7] P. Chen, Z. Xiong, J. Luo, J. Lin and K.-L. Tan, *J. Phys. Chem.*, B 107, 2003, p 10967-10970.

[8] T. Ichikawa, N. Hanada, S. Isobe, H. Leng and H. Fujii, *J. Phys. Chem.*, B108, 2004, p 7887-7892

[9] Y. Hang and E. Ruckenstein, *Ind. Eng. Chem. Res.*, 44, 2005, p 1510-1513.

[10] A. Huq, J.W. Richardson Jr., E. Maxey, D. Chandra, and W. Chien, *J. Alloys Compd.*, 436, 2007, p 256-260.

[11] A. Huq, J.W. Richardson Jr., E. Maxey, D. Chandra, and W. Chien, *J. Phys. Chem. C*, 111 (28), 2007, p 10712-10717.

[12] W. Chien, J. Lamb, D. Chandra, A. Huq, J. Richardson Jr., and E. Maxey, *J. Alloys Compd.*, 446-447, 2007, p 363-367.

(368b) Hydrogen Absorption-Desorption Cycle Durability and CO Tolerance of Hydrogen Absorbing Alloys Developed in WE-NET Project

Hideaki Tanaka[1], Kenji Aihara[2], Tomoyuki Miyata[2], Naoki Taoka[2], Kozo Mori[2], Teruo Kaneko[2] and **Nobuhiro Kuriyama**[1], (1)Research Institute for Ubiquitous Energy Devices, National Institute of Advanced Industrial Science and Technology, Ikeda, Japan, (2)New Material Center, Osaka Science and Technology Center, Osaka, Japan

Hydrogen absorbing alloys, "traditional alloys", have been studied for the applications to nickel-metal hydride batteries and hydrogen storage tanks for portable power generators and fuel cell vehicles, because of their high volumetric hydrogen density over 90 kg-H2/m3 and good kinetics for the reaction with hydrogen. IMURA, Japan Metals & Chemicals, Japan Steel Works, Sumitomo Steel, and OSTEC had promoted development of high-capacity alloys for room-temperature use in the second phase (FY1999-FY2002) of WE-NET Project, the Japanese national project for hydrogen systems, and Ca-Mg-Ni-based Laves-phase alloys, for example Ca0.25Mg0.66Y0.09Ni1.86, and BCC-structured alloys, Ti1.01Cr1.49V0.30, Ti1.10Cr1.39Mo0.30, V74.5Ti10Cr12.5Mn3, and V75.5Ti10Cr11Mn2Ni1.5. We will report hydrogen absorption-desorption cycle durability and influence of CO impurity of the novel alloys in order to give information for development of hydrogen storage systems. Although Ca0.25Mg0.66Y0.09Ni1.86 indicated considerable decrease, ca 40%, in hydrogen capacity within initial 20 cycles, decrease in prolonged cycles from 20th cycle to 400th was only 10% of the first cycle. This alloy indicated considerable influence of CO in hydrogen even at 1 ppm. The BCC alloys showed stable capacity over 1000 cycles in pure hydrogen and hydrogen containing 1 ppm of CO, although the capacities of Ti1.01Cr1.49V0.30 and Ti1.10Cr1.39Mo0.30 deceased ca. 15% within initial 100 cycles. Those alloy indicated appreciable influence by 10 ppm of CO in hydrogen.

See more of this Session: Hydrogen Storage System Engineering and Applications: Poster Session
See more of this Group/Topical: Topical 8: Hydrogen Production and Storage

(368c) Hydrogen Storage System Using $LiBH_4$

Hiroyuki T. Takeshita, Toyokazu Eguchi, Masanori Hayakawa and Makoto Ikeda, Department of Chemistry and Materials Engineering, Kansai university, Suita, Osaka, Japan

$LiBH_4$ is a promising candidate for hydrogen storage due to its high gravimetric hydrogen capacity, but improvement must be achieved for thermodynamic stability, slow reaction kinetics and reactivity against impurity gas species. We propose that thermodynamic and kinetic problems can be improved by using the mixture composed of $LiBH_4$, MgH_2 and Al and that reactivity against impurity gas can be improved by utilization of excess $LiBH_4$ which reacts with impurity gas species such as water vapor and works as hydrogen source as a result. In this paper, firstly the mixture of $4LiBH_4$-MgH_2-xAl (x=1 or less) is reported on its constituent phases, kinetic hydrogen desorption properties and cyclic de-/re-hydrogenation properties, and then $LiBH_4$ is reported on its reaction against impurity gas species, for example H_2O and O_2, and its hydrogen emission in heating after poisoning.

See more of this Session: Hydrogen Storage System Engineering and Applications: Poster Session
See more of this Group/Topical: Topical 8: Hydrogen Production and Storage

Hydrogen Purification and Storage System Using Metal Hydride

Shinichi Miura[1], Akitoshi Fujisawa[1] and Masayoshi Ishida[2], (1)Kobe Steel,Ltd., Kobe, Japan, (2)University of Tsukuba, Tsukuba, Japan

This work is to develop a new hydrogen purification and storage system for daily start and stop operations. The new system enables us to minimize emissions of carbon dioxide by using compact and highly efficient fuel cells.

As a result of a joint investigation conducted by Kobe Steel and the University of Tsukuba, a new process is proposed. First, this method thoroughly eliminates carbon monoxide using an adsorption material that strongly selects carbon monoxide, and second it introduces a mixed gas to a metal hydride that purifies and stores hydrogen. Thus, the COA-MIB (CO Adsorption Metal Hydride Intermediate- Buffer) method was established.

The selective adsorbent of carbon monoxide used in this process is an adsorbent based on porous alumina supporting Cu (I). Chemical adsorption of carbon monoxide occurs on the supported Cu (I). If carbon monoxide exists in reformed gas, carbon monoxide is adsorbed on the surface of metal hydride and it prevents hydrogen from being absorbed. It is necessary to eliminate carbon monoxide to achieve a hydrogen purification process using a metal hydride. After removing carbon monoxide, the reformed gas is introduced into a metal hydride bed to purify and store hydrogen.

Metal hydride has a high storage volume efficiency. If no poisonous substances are included, metal hydride can be used as a gas purification and storage material that selectively stores hydrogen when a mixed gas is supplied to it. As far as a fixed-type of apparatus is concerned, the existing AB5 type of metal hydride can be used.

Metal hydride absorbs and releases hydrogen in an equilibrium process according to the designed alloy temperature and pressure. When it is maintained at an equilibrium temperature it will not affect the hydrogen maintaining pressure, and it keeps the absorbed amount of hydrogen. On the other hand, when the pressure in a system falls because hydrogen is extracted, the metal hydride releases hydrogen in order to keep the equilibrium pressure. By making the best of this characteristic, a power generation system with a passive adjustment of supply and demand can be realized if a fuel cell is connected to the metal hydride that stores pure and pressurized hydrogen. In other words, a system can be made where the consumed hydrogen is compensated for by the metal hydride.

Some 100 NL/h Laboratory scale apparatus was designed and operated in daily start and stop operations for 100 cycles for a total of 150 hours with quite good efficiency. The new process has achieved an 83% hydrogen recovery ratio in one-month DSS operations.

(403a) Optimization of Hydrogen Storage Tanks: The Case of Sodium Alanate

Gustavo A. Lozano[1], Chakkrit Na Ranong[2], José M. Bellosta von Colbe[1], Rüdiger Bormann[1], Jobst Hapke[2], Georg Fieg[2], Thomas Klassen[1] and Martin Dornheim[1], (1)Institute of Materials Research, Materials Technology, GKSS Research Centre Geesthacht, Geesthacht, Germany, (2)Institute of Process and Plant Engineering, Hamburg University of Technology, Hamburg

Hydrogen is a very promising energy carrier for a comprehensive clean-energy concept in mobile applications. Regarding its use as fuel for the zero-emission vehicle, one main challenge is its storage. Hydrogen storage systems should fulfil the requirements of automotive applications, i.e. high gravimetric and volumetric storage densities, fast charging and discharging rates at moderate operating conditions, and high safety levels. Solid storage of hydrogen as metal hydrides offers a safe alternative to hydrogen storage in compressed or liquid form and has higher volumetric storage density. The light metal hydride sodium alanate, $NaAlH_4$, compared to classical room temperature hydrides, offers a suitable compromise with relatively large gravimetric storage densities at rather moderate operating pressures and temperatures. Design of practical hydrogen storage systems aims at minimizing volume and weight while fulfilling defined criteria such as driving ranges and storage densities of the complete on-board tank system.

In the present investigation, hydrogen storage tanks based on sodium alanate are experimentally investigated, mathematically modelled, simulated and finally optimized. The determination of an optimal tubular storage tank based on sodium alanate material is presented. The optimisation is carried out on the basis of the predictions of a developed simulation tool for the hydrogen absorption, supported on the experimental results. Optimisation is performed for tubular tanks filled with loose powder and with compacted material. Compared to loose powder compacted material has the advantage of higher volumetric storage density and higher thermal conductivity, but the disadvantage of lower permeability to hydrogen flow and higher volumetric heat release during hydrogenation. Beyond compaction of the hydride material, the addition of expanded graphite is also evaluated in the optimisations. Although increasing the inert material of the system and thus reducing the total hydrogen storage capacity, the addition of expanded graphite enhances the effective thermal conductivity of the hydride bed.

See more of this Session: Hydrogen Storage System Engineering and Applications: Heat and Mass Transfer Modeling
See more of this Group/Topical: Topical 8: Hydrogen Production and Storage

(403b) Modeling of Adsorbent Based Hydrogen Storage Systems

Bruce Hardy[1], Claudio Corgnale[2], Richard Chahine[3], Marc-André Richard[4], David A. Tamburello[1], Stephen Garrison[1] and Donald Anton[5], (1)Computational Sciences, Savannah River National Laboratory, Aiken, SC, (2)Hydrogen Technology Center, Savannah River National Laboratory, Aiken, SC, (3)Hydrogen Research Institute, Trois-Rivieres, QC, Canada, (4)Hydro-Quebec, (5)Energy Security, Savannah River National Laboratory, Aiken, SC

Adsorbent based hydrogen storage systems have high potential for meeting the DOE Technical System Targets for light vehicles. However, these systems must charge and be maintained at low temperatures to retain a useable quantity of hydrogen. Heat and mass transfer considerations for these low temperature systems, together with the thermodynamics of the adsorbent media, results in complex behavior that must be addressed for application to on-board vehicular storage. In general, the dynamic behavior of media based hydrogen storage systems is sufficiently complex that numerical models are required for evaluation and design. Numerical models suitable for engineering analyses of media based storage systems must couple the mass, momentum and energy conservation equations to the chemical kinetics equations, or isotherms, for the media as appropriate. Typically, for storage media having very rapid kinetics, state dependent models for isotherms along with mass transfer resistances are used to determine the amount and rate of hydrogen uptake or discharge.

This paper describes the application of numerical models to the evaluation of hydrogen storage systems that utilize MaxSorb MSC-30™ and MOF-5™ adsorbents. The models for these systems incorporate thermodynamic parameters obtained by fitting data to the Dubinin-Astakhov adsorption model. Non-ideal gas properties for hydrogen are required for the range of pressures and temperatures over which these systems operate. The models can be applied in up to 3 dimensions for arbitrary vessel geometries and boundary conditions. Validation is performed against experimental data for MaxSorb MSC-30™.

Designs for temperature control using heat transfer surfaces of various geometries and flow-through cooling by hydrogen gas are investigated. Effects of cooling methods and media compaction on charging rates, capacity, and system performance are assessed. High pressure systems, suitable for hybrid storage, in which a significant amount of hydrogen resides in inter-particle void spaces and pores, as well as in the adsorption volume, are considered.

See more of this Session: Hydrogen Storage System Engineering and Applications: Heat and Mass Transfer Modeling
See more of this Group/Topical: Topical 8: Hydrogen Production and Storage

(403c) Model Based Design of An Automotive-Scale, Metal Hydride Hydrogen Storage System

Terry A. Johnson[1], Daniel E. Dedrick[1], Michael P. Kanouff[1], Gregory H. Evans[1] and Scott W. Jorgensen[2], (1)Thermal Fluids Science and Engineering, Sandia National Labs, Livermore, CA, (2) General Motors Research and Development, Warren, MI

Sandia and General Motors have successfully designed, fabricated, and experimentally operated a vehicle-scale hydrogen storage system using the complex metal hydride sodium alanate. Over the 6 year project, the team tackled the primary barriers associated with storage and delivery of hydrogen including mass, volume, efficiency and cost. The result was the hydrogen storage demonstration system design.

The key technologies developed for this hydrogen storage system include optimal heat exchange designs, thermal properties enhancement, a unique catalytic hydrogen burner and energy efficient control schemes. The prototype system designed, built, and operated to demonstrate these technologies consists of four identical hydrogen storage modules with a total hydrogen capacity of 3 kg. Each module consists of twelve stainless steel tubes that contain the enhanced sodium alanate. The tubes are arranged in a staggered, 4 x 3 array and enclosed by a steel shell to form a shell and tube heat exchanger. Temperature control during hydrogen absorption and desorption is accomplished by circulating a heat transfer fluid through each module shell. For desorption, heat is provided by the catalytic oxidation of hydrogen within a high efficiency, compact heat exchanger. The heater was designed to transfer up to 30 kW of heat from the catalytic reaction to the circulating heat transfer fluid.

The demonstration system module design and the system control strategies were enabled by experiment-based, computational simulations that included heat and mass transfer coupled with chemical kinetics. Module heat exchange systems were optimized using multi-dimensional models of coupled fluid dynamics and heat transfer. Chemical kinetics models were coupled with both heat and mass transfer calculations to design the sodium alanate vessels. Fluid flow distribution was a key aspect of the design for the hydrogen storage modules and computational simulations were used to balance heat transfer with fluid pressure requirements.

An overview of the hydrogen storage system will be given, and examples of these models and simulation results will be described and related to component design. In addition, comparisons of demonstration system experimental results to model predictions will be reported.

See more of this Session: Hydrogen Storage System Engineering and Applications: Heat and Mass Transfer Modeling
See more of this Group/Topical: Topical 8: Hydrogen Production and Storage

(403d) Scaled up LiBH4 - MgH2 Composite Storage System as New Promising Hydrogen Storage

Julian Jepsen, José Bellosta von Colbe, Ulrike Bösenberg, Gustavo Lozano, Thomas Klassen and Martin Dornheim, Institute of Materials Research, Materials Technology, GKSS-Forschungszentrum Geesthacht GmbH, Geesthacht, Germany

The discovery of the reaction mechanism of the Reactive Hydride Composite (RHC) some years ago, where the reaction of two metal hydrides lowers the overall reaction enthalpy [1, 2] gave new impetus to the entire hydrogen storage technology. RHCs combine high hydrogen storage capacities and relatively fast sorption rates, but at more moderate temperature and pressure conditions compared to the conventional high pressure (up to 700 bar) and liquid (-253 °C) storage systems. In particular for the promising combination of LiBH4 and MgH2 many investigations were done for a deep understanding of the reaction pathways (e.g. [3]). However, a scaled up tank system containing the LiBH4-MgH2 composite had not been investigated until now. In this paper we present the first results of such a system concerning the storage capacity, sorption rates and cycling behaviour. The investigated maximum temperature and hydrogen pressure conditions are at 400 °C and 50 bar respectively. The inner tank volume is 600 cm3. The theoretical investigation is performed by calculation of the mass and energy balances. Besides the use of a fluid heat transfer medium we also investigate the possibility of a electrical heat management. Hence we achieve further considerable weight and volume savings. However, the main advantage is the possibility of a free form tank system due to a flexible heat management and the maximum pressure of 50 bar. A free form tank system is advantageous for the automotive industry and most other mobile and stationary applications and can be considered as breakthrough in the hydrogen storage technology for high temperature hydrides.

References [1] Barkhordarian, G., Klassen, T., Dornheim, M., Bormann, R., 2007. Unexpected kinetic effect of MgB2 in reactive hydride composites containing complex borohydrides. Journal of Alloys and Compounds 440, L18-L21. [2] Vajo, J.J., Skeith, S.L., Mertens, F., 2005. Reversible Storage of Hydrogen in Destabilized LiBH4. J. Phys. Chem. B 109, 3719-3722. [3] Bösenberg, U., Doppiu, S., Mosegaard, L., Barkhordarian, G., Eigen, N., Borgschulte, A., Jensen, T.R., Cerenius, Y., Gutfleisch, O., Klassen, T., Dornheim, M., Bormann, R., 2007. Hydrogen sorption properties of MgH2-LiBH4 composites. Acta Materialia 55, 3951-3958.

See more of this Session: Hydrogen Storage System Engineering and Applications: Heat and Mass Transfer Modeling
See more of this Group/Topical: Topical 8: Hydrogen Production and Storage

(403e) Evaluation of Acceptability Envelope for Materials-Based H2 Storage Systems

Claudio Corgnale[1], Bruce Hardy[2], David A. Tamburello[2], Stephen L. Garrison[3] and Donald Anton[4], (1) Hydrogen Technology Center, Savannah River National Laboratory, Aiken, SC, (2)Computational Sciences, Savannah River National Laboratory, Aiken, SC, (3)Savannah River National Laboratory, Aiken, SC, (4)Energy Security, Savannah River National Laboratory, Aiken, SC

One of the most important issues to be addressed to make hydrogen a viable energy carrier is its storage. Among other possibilities (liquid storage, gaseous storage) solid state H2 storage systems, by means of adequate materials, seem to be one of the most suitable solutions being able to achieve high gravimetric and volumetric capacities. The paper shows results of activities carried out at Savannah River National Laboratory to evaluate acceptability envelope for such systems. The aim of the acceptability envelope is to determine the properties of the coupled media and system needed to meet storage technical specifications. To do that, a steady state model has been set up, capable to simulate the behavior of different materials under different operating conditions. Various scenarios have been selected, starting from the DOE technical target specifications, and different H2 storage materials have been analyzed and compared. In addition, several sensitivity analyses have been carried out showing the influence of selected thermo-physical material quantities on characteristic sizes of the storage system.

See more of this Session: Hydrogen Storage System Engineering and Applications: Heat and Mass Transfer Modeling
See more of this Group/Topical: Topical 8: Hydrogen Production and Storage

(403f) Automatic Optimization of Metal Hydride Storage Tanks and Novel Designs

Stephen L. Garrison[1], Mikhail B. Gorbounov[2], David A. Tamburello[3], Bruce Hardy[3], Claudio Corgnale[4], Daniel A. Mosher[5] and Donald Anton[6], (1)Savannah River National Laboratory, Aiken, SC, (2)Thermal Management, United Technologies Research Center, East Hartford, CT, (3)Computational Sciences, Savannah River National Laboratory, Aiken, SC, (4)Hydrogen Technology Center, Savannah River National Laboratory, Aiken, SC, (5)Applied Mechanics, United Technologies Research Center, East Hartford, CT, (6)Energy Security, Savannah River National Laboratory, Aiken, SC

A key technical hurdle to an energy economy focused on clean-burning hydrogen over non-renewable petroleum is on-board storage of hydrogen for automotive vehicular applications. Metal hydrides provide one possible solution to the need for high gravimetric and volumetric capacities. However, metal hydrides release significant amounts of heat during hydrogen adsorption, on the order of 0.1 GJ. The subsequent rise in storage temperature results in decreases in maximum hydrogen storage capacity and loading rate at the short fill times mandated by the Department of Energy (DOE) targets. Integrated tank heat exchangers, e.g., cooling tubes and fins, are needed to maximize heat transfer out of system to maximize storage rate and capacity. Given the high cost of experiments (especially at higher pressures), the non-commodity nature of some metal hydrides, and the possibility of air- or water- reactivity, computational modeling provides a significant contribution to a comprehensive methodology for the design, evaluation, and modification of hydrogen storage systems. While many literature studies focus on maximum capacity assuming an (impossible) infinite fill time, full, detailed modeling of the coupled heat and mass transfer and chemical reaction kinetics is needed. The focus of this presentation will be our work developing detailed numerical models that couple reaction kinetics with heat and mass transfer via the general purpose finite element solver COMSOL Multiphysics® for general metal hydride beds and the automatic optimization of the designs by linking the COMSOL models with routines in MATLAB®. Optimization of novel heat exchanger designs will also be presented.

See more of this Session: Hydrogen Storage System Engineering and Applications: Heat and Mass Transfer Modeling
See more of this Group/Topical: Topical 8: Hydrogen Production and Storage

(427a) Liquid Phase Reforming of Wood Flour to Hydrogen

Timothy D. Davis, Sean C. Emerson, Tianli Zhu, Rhonda R. Willigan, Ying She and Thomas Henry Vanderspurt, United Technologies Research Center, East Hartford, CT

Background

Growing concern over the availability, cost, and environmental impact of traditional petroleum energy sources, has emphasized the need to develop alternative energy technologies that reduce our dependence on non-renewable resources. Lignocellulosic biomass is one such low cost source that is renewable and available in significant quantities. Globally, over 224 billion dry tones of biomass are produced through photosynthesis each year, roughly 10 times the world's energy consumption[1]. Currently, significant levels of research are in the field of converting this sustainable feedstock into many forms of energy, including syngas and several forms of liquid fuels.

Hydrogen is another possible product from biomass. H_2 is useful for numerous applications ranging from fuel cells, chemical processing, or direct consumption as a fuel. The challenge that remains is to provide a low cost, efficient route for its conversion to hydrogen. For such a technology to be economically feasible, a proposed system must be simple, using raw biomass feedstock, have a high selectivity for H_2 production, and not produce toxic byproducts. Significant work has been reported in the literature for liquid phase reforming (LPR) of preprocessed biomass derived feed stocks and related oxygenates like ethylene glycol. Solutions of glycerol, glucose or sorbitol have been reformed to H_2, CO_2, and low molecular weight alkanes over Pt, Ni-Sn catalysts[2,3] and other[4] catalysts. This work concentrates on the production of H2 directly from raw biomass, without the need for intermediate processing steps.

In 2009 we reported on the high efficiency Liquid Phase Reforming of yellow poplar (***Liriodendron tulipifera)*** saw dust over Pt-Re/ceria zirconia catalyst pellets. The striking observation was that the effluent was water white and substantially free of organic matter except for a trace amounts of formic and possibly acetic acid[5]. Thus not only were the hemi-cellulose and cellulose reformed but the lignin as well. This catalyst was selective towards hydrogen co-producing relatively little methane and higher alkanes. It was also robust, but in practice this catalyst would be expensive. Furthermore the high surface area–large pore cubic ceria-zirconia support was not readily available as an item of commerce. Thus one of the next steps before proceeding to a continuous process converting wood to hydrogen was the identification and kinetic characterization of a less expensive, more readily available catalyst.

Materials and Methods

The biomass materials and surrogates used in this work include ethylene glycol and 200 mesh wood flour obtained from P. J. Murphy. The catalyst starting materials was Raney Ni obtained from W. R. Grace. Kinetic experiments were performed in a custom build 0.5" OD inch flow reactor constructed of Inconel 625. The system was heated by means of an infrared furnace with an Isco pump, Model 500D, supplying the feed to the reactor. A catalytic bed bed was placed in the reactor, supported by ceria stabilized zirconia beads. Pressure was controlled by means of a Tescom back pressure regulator equipped with an ER3000 electronic pressure controller. Gas analysis was performed by means of dual Varian 3800 GCs, for hydrocarbon and permanent gas analysis.

Results and Discussion

Kinetic studies on the liquid phase reforming of a biomass surrogate, ethylene glycol, have been carried out in alkaline solutions over several base metal catalysts. Several bases and concentrations have been studied to investigate the effect of basicity and concentration on the system. Although not required for the LPR, the alkalis are for the hydrolysis of the biomass. Of the metals, the study was quickly down selected to Raney Nickel, as it was concluded that the other catalysts tested did not have sufficient C-C bond breaking ability at low temperatures to warrant further study. The complete conversion of wood flour to gas and water white carbonate solution was demonstrated with Raney Ni was confirmed in batch reactor studies.

Flow reactor studies of the LPR of 5 wt.% ethylene glycol over an 8g bed of Raney nickel were performed. Selectivity and yield results are given in Figures 1 and 2. As expected, Raney nickel was capable of completely reforming ethylene glycol to gaseous species at 310 °C in water. In addition, a high level of methanation was evident, an inherent characteristic of Ni reforming catalysts. Increasing amounts of K_2CO_3 slightly decreased the methanation activity as well as reduced the overall yields. However, increasing concentrations of KOH in the system did not detrimentally affect the yield but dramatically decreased the methanation activity, resulting in significantly higher selectivity's. This reduction of methanation activity may be due to homogeneously catalyzed water gas shift reactions, driving any formed carbon monoxide to carbon dioxide before methane can be produced. CO_2 methanation requires higher temperatures to methanate in any significant quantity.

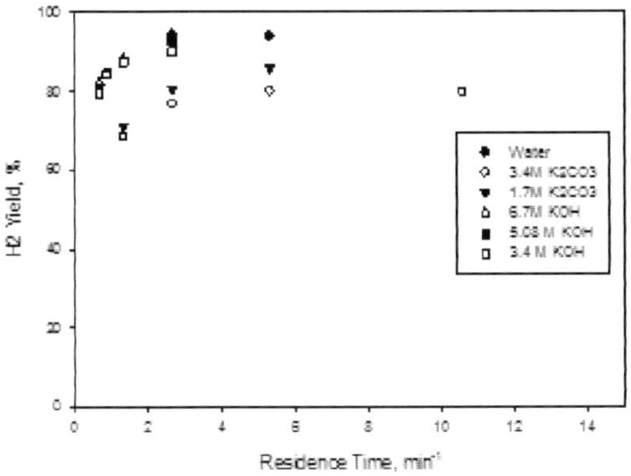

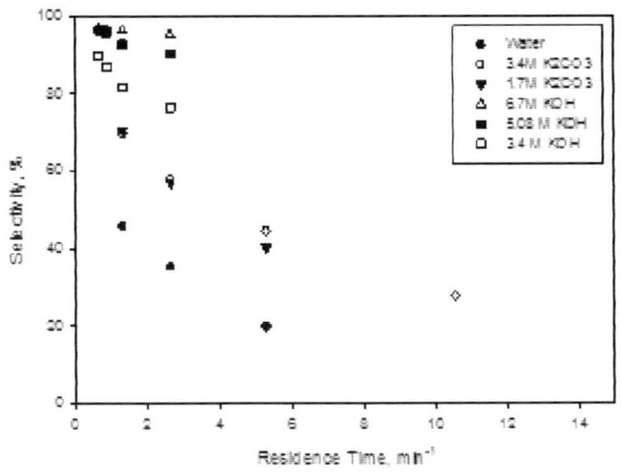

Figures 1 and 2- LPR of 5 wt.% Ethylene Glycol over Raney Nickel, 310 °C, 120 atm.

However, using high levels of hydroxide to improve the selectivity of nickel based catalysts creates a system level energetic issue. The KOH is converted to K_2CO_3 in situ, and would require regeneration. As an alternative, a study was performed of doping the Raney nickel catalyst to promoters to drive down the methanation activity. As shown in Figures 3, a small addition of a promoter was sufficient to drive down the methanation activity without significantly decreasing overall reforming activity. This result enables the use of K_2CO_3 as the base for the wood hydrolysis, which has already been demonstrated as sufficient for the process[5].

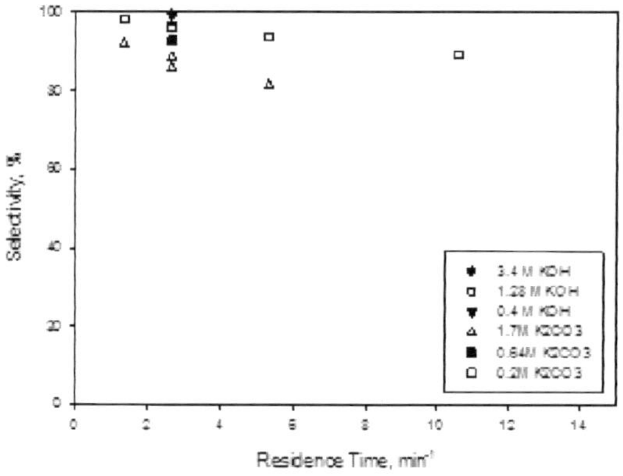

Figures 3- LPR of 2.5 wt.% Ethylene Glycol over Promoted Raney Nickel, 310 °C, 120 atm.

Work is ongoing to understand the LPR mechanism. A combination of experimentation and atomistic modeling will help explain the mechanism. This may lead to catalyst modification enabling faster kinetics and tailoring the selectivity to meet the plant system efficiency targets. Furthermore, the surrogate results will be compared to those obtained from directly pumping and reacting raw biomass into the reactor.

In order to bring the technology one step closer to realization, a prototype power plant will be constructed to demonstrate the feasibility of the system for processing of raw biomass. The system is targeted to produce hydrogen sufficient to power a 1 kWe Fuel Cell by liquid phase reforming of wood flour. As of the time of this abstract, a high pressure slurry pump has been acquired and validated for pumping the feedstock at the desired conditions. Work is ongoing to determine the kinetics and optimal conditions for the wood hydrolysis. The system will use a promoted base metal catalyst for reforming and a palladium membrane for the separation of pure hydrogen from the reformate.

Significance

This work moves the concept of hydrogen production from raw biomass as describe in DOE contract DE-FG36-05GO15042 "A Novel Slurry-Based Biomass Reforming Process" a significant step closer to realization.

See more of this Session: Renewable Hydrogen Production I
See more of this Group/Topical: Topical 8: Hydrogen Production and Storage

NICKEL CATALYZED SUBCRITICAL AQUEOUS PHASE LIQUEFACTION OF BIOMASS FOR HYDROGEN PRODUCTION

Richa Tungal[1], *Rajesh V. Shende*[1], *Lew Christopher*[2]

[1]Department of Chemical and Biological Engineering,
[2]Center for Bioprocessing Research & Development,
South Dakota School of Mines & Technology,
Rapid City, SD 57701 USA

Abstract

This paper reports subcritical aqueous phase liquefaction of office paper waste in presence of homogeneous nickel catalysts for H_2 production. Slurry of 10-20 wt% biomass was prepared in distilled water containing 5 wt% Ni-catalyst and loaded in high-temperature high-pressure PARR reactor and heated to 250°C. The reaction progress was monitored for 2 h at 250°C and during course of reaction; gas samples were withdrawn at regular time intervals and analyzed using Chrompak capillary column on the GC equipped with thermal conductivity detector. The analysis of gas samples indicated the presence of H_2, CO_2, CO, and CH_4. Different Ni-salts that were used as catalyst during liquefaction of the biomass were found to have significant effect on H_2 generation. Specifically, after 1.5 h of reaction, $NiSO_4$, $NiCl_2$ and $Ni(NO_3)_2$ produced H_2 in the vol% of 18%, 40% and 43%, respectively. The effectiveness of a Ni-catalyst on H_2 generation from subcritical hydrothermal processing of cellulose was also investigated as it is the major component present in commonly used office paper.

Introduction

Almost every office around the world uses some kind of paper for multiple reasons. The unwanted used papers are then directly disposed off or shredded and disposed. U.S. produces about 4 million tons of office paper annually, whereas worldwide more than 300 million tons of different variety of papers are manufactured. It has been estimated that for the production of one ton of paper, processing of about 2 - 4 tons of trees are required. Thus significant biomass processing is required for paper production. As the usage of paper and paper products has been increasing day by day, the production of paper waste is also increasing, creating paper waste disposable issues [1]. The production of municipal solid waste in U.S alone is about 3200 pounds annually and out of that about 1500 pounds is a paper waste [2]. Disposing off this large amount of paper waste is posing a great problem due to high costs involved. Currently, most of the paper waste is being discarded through landfill and incineration and only a small portion is being recycled. Worldwide about 30 - 50% of waste paper is disposed by landfilling, majority paper waste is incinerated and only about 10% is recycled. Many industrialized countries recycle paper waste but the recycling efforts vary greatly with different countries, for instance Germany and India recycle about 80% and 20% respectively. As the recycled paper has high content of pulp fiber, it is difficult to get new paper products from recycling [3]. Moreover, due to heavy recycling cost, the market of the recycled paper is limited.

From waste management point of view, it appears that additional energy and efforts are required for the disposal of paper waste. Instead, if energy or fuel is derived from the paper waste, it will open new avenues for energy production as well as waste reduction. There exist thermochemical and biochemical or biological technologies for converting paper waste into gaseous or liquid fuels, which include H_2, lower hydrocarbons and alcohols. Among these technologies, one of the thermochemical processes such as high pressure hydrothermal treatment of waste biomass is attracting great deal of attention [4]. High pressure hydrothermal processes can be either subcritical, where water is operated below critical temperature (T_c) and pressure (P_c) of 374.14°C and 22.064 MPa or supercritical, if water is used above the critical temperature and pressure conditions. Subcritical hydrothermal process has advantages over other thermochemical processes that it can directly process and liquefy wet waste biomass at low temperatures and saves energy from drying operation. This process can be operated under gasification or pyrolysis conditions for liquefaction of a biomass and production of valuable fuels. Recently, rapid conversion of switchgrass in subcritical water under pyrolysis conditions was performed at temperatures of 250- 350°C and pressures upto 20 MPa and this study reported significant presence of H_2 in the gaseous products [5]. In another study, paper waste was processed under subcritical water conditions in presence of charcoal supported ruthenium catalyst at 250-300°C and reported production of methane, H_2 and other gases. In this study, however, very small amount of H_2 generation has been indicated [6]. Many other investigations reported production of different fuels under various hydrothermal conditions.

Among different fuels currently available, H_2 is a green fuel with very high energy density. H_2 can be produced from lignocellulosic biomass using thermal, biological and chemical processes. As U.S is currently producing about 400,000 tons/year of lignocellulosic biomass, there is significant potential exists for H_2 production [7]. For the thermochemical processes, however a great deal of research and development efforts are needed in order for this technology to be cost-competitive over other conventionally used H_2 production technologies. At present R&D activities are focused on the development of the catalytic processes with the goal of achieving highest possible H_2 levels from a biomass processing at moderate temperature and pressure conditions. Therefore, we believe it will be very beneficial if significant level of H_2 is obtained at subcritical hydrothermal processing conditions in presence of suitable catalysts. Few investigators have already performed hydrothermal gasification of glucose in presence of raney nickel and several other organometallic catalysts at 310-350°C and reported higher H_2 levels as compared with non-catalytic conditions [8]. In another study, cellulose was gasified to H_2 rich gas using reduced nickel catalyst at hydrothermal conditions of 350°C and 18 MPa [9]. Antal et al. used glucose as potential feedstock for H_2 production via steam reforming in presence of different metal oxide supported metal (Ni/SiO_2) catalyst and reported H_2 rich syn gas [10]. Most of these studies were focused on H_2 generation from model compounds. However, very little effort is noticed in the literature on the production of H_2 from paper waste.

In this paper subcritical hydrothermal processing of office paper waste was performed using three different homogeneous Ni-catalysts at 250°C. The reaction was monitored for 2 h and it was in progress, the gas samples were withdrawn and analyzed using GC. The results obtained on H_2 volume generation are presented and the effectiveness of catalysts is discussed.

Experimental Methods and Analysis

Materials

$NiSO_4.6H_2O$, $NiCl_2.6H_2O$ and $Ni(NO_3)_2.6H_2O$ were purchased from Alfa Aesar and and used without purification. α-Cellulose was purchased from Sigma Aldrich, WI. N_2 from a cylinder with the minimum stated purity of 99% was purchased from Linweld Inc., SD. The shredded office paper waste or cellulose was made into slurry in distilled water and used for subcritical hydrothermal processing in N_2 environment.

Experimental Set-up and Procedure

High-temperature high-pressure PARR SS316 reactor of 300 ml capacity equipped with heating and cooling assembly, gas inlet and outlet valves, sampling valves, pressure gauge, safety rupture disc, thermocouple, magnetic stirrer and temperature indicator controller was used to perform subcritical hydrothermal processing of a biomass. This reactor is rated for maximum working temperature and pressure of 350°C and 3000 psi, respectively. An additional SS gas sampling vessel was mounted on the reactor. The entire reactor set-up is shown below in Figure 1.

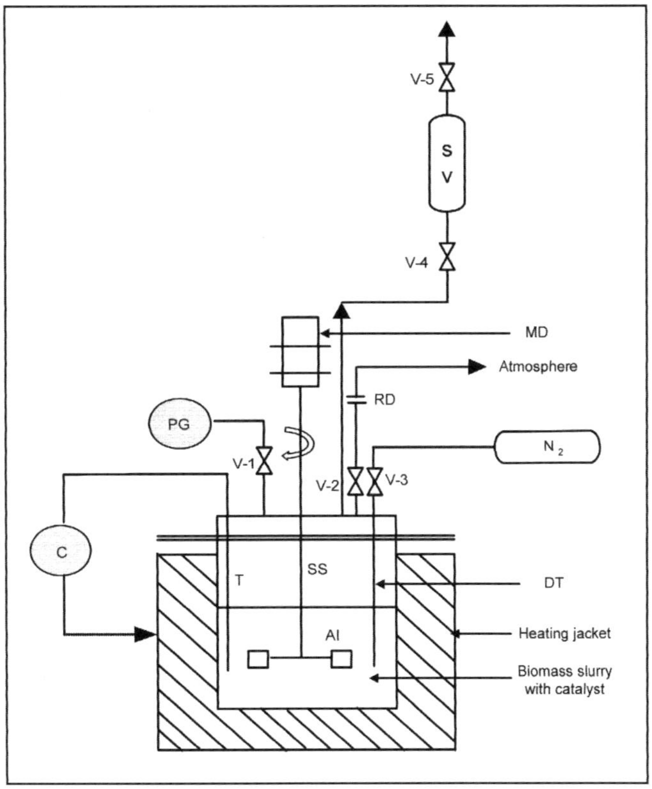

Fig. 1: Reactor set-up for subcritical hydropyrolysis of a biomass for H_2 generation. (RD-rupture disc, T-thermocouple, AI-adjustable impeller, SS-stirring shaft, PG-pressure gauge, MD-magnetic drive, V-valve, DT-dip tube, C-controller, SV-gas sampling vessel).

Accurately weighed 15.0 g of shredded office paper waste or cellulose was made into slurry in 150 ml distilled water containing 5wt% (taken on a biomass weight basis) Ni-catalyst and loaded in the high-temperature high-pressure PARR reactor. The contents in the reactor were flushed continuously with N_2 for 10-15 min to replace air and pressurized to 40 psi. The reactor was heated to 250°C under continuous magnetic stirring of a biomass slurry at 1300 rpm. It took approx. 45 min to attain the set point of 250°C where the total reactor pressure was 700 psi. Once the reactor temperature was attained, a gas sample was withdrawn. This time was taken as 'zero' time for the reaction. The reaction was continued for 120 minutes and during the course of the reaction, gas samples were withdrawn at regular time intervals and analyzed on the GC. At the end of 120 min, the reactor contents were cooled down to room temperature and the liquid obtained was filtered to determine the amount of residue.

GC Analysis and Residue Measurement

Gas samples were withdrawn with a gas tight syringe during the subcritical hydrothermal processing of office paper waste or cellulose and analyzed using Agilent 7890A GC equipped with Chrompak capillary column and thermal conductivity detector. A sample injection volume of 100 μl was used for the analysis. The GC analysis was performed under the isothermal conditions where column and detector temperatures were maintained at 150°C and 205°C, respectively.

At the end of the reaction, the reactor contents were cooled down to room temperature and filtered using Whatman filter paper. The residue was recovered after filtration and dried in the oven at 60°C and weighed accurately.

Results and Discussion

Figure 2 shows the product gas profile obtained during subcritical hydropyrolysis of office paper waste at 250°C in presence of $NiCl_2$ catalyst. The product gas profile indicates primarily H_2, CO, CH_4 and CO_2 gases. In addition to these gases, H_2O vapor should also be present in the gas samples, however, as the gas sampling vessel was held at ambient conditions, the water vapors might have been already condensed. Moreover the Chrompak capillary column that was used for the analysis of product gases was not capable of detecting water vapors. Therefore, the concentrations were determined without accounting for H_2O vapor in terms of vol% of total headspace volume, which was 150 ml. The product gas profile presented in Figure 2 indicates significant concentration levels of H_2 and CO_2 gases followed by lower level of CO, however, negligible volume of CH_4. The figure also indicates that as the reaction time increases, the concentration of H_2 and CO_2 increases and the maximum H_2 generation is approx. 40 vol% at 90 min of reaction time. It is to be noted that in absence of $NiCl_2$ catalyst, no H_2 generation was observed from the office paper waste under the subcritical hydrothermal processing at 250°C. Therefore from the experiments performed we believe that the H_2 generated was solely due to the catalytic action of $NiCl_2$ and that the reactor wall effects were negligible. Our finding is consistent with the study where investigators [11] had reported significant H_2 production in presence of $NiCl_2$ and other metal salts catalyst.

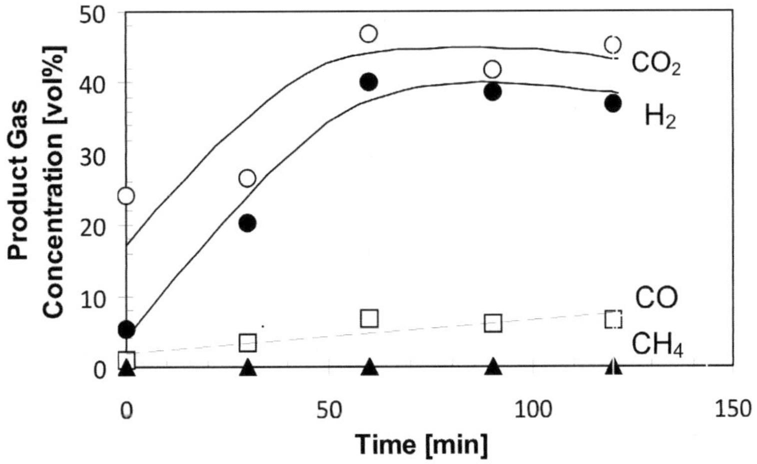

Fig. 2: Product gas profile showing H_2, CH_4, CO and CO_2 gases during subcritical hydropyrolysis of office waste paper at 250°C in presence of 5 wt % $NiCl_2$.

The product gas profile obtained during subcritical hydrothermal processing of office paper waste at 250°C in presence of $Ni(NO_3)_2$ catalyst is presented in Figure 3. Analogous to $NiCl_2$ effect on the product gas profile, the $Ni(NO_3)_2$ catalyst indicated the presence of H_2, CO_2, CO and CH_4 gases. These concentration profiles further reveal H_2 and CO_2 generation in higher volumes as compared with CO; however, CH_4 volume was in negligible amount. The profiles also indicate almost no change in CO concentration. Further observation shows that the H_2 concentration increases to the maximum value of about 43vol% that corresponds to the reaction time of 90 min and further decreases slightly. The CO_2 concentration levels observed with $Ni(NO_3)_2$ catalyst were higher by about 20% than those observed with $NiCl_2$ (Figure 2). This suggests that $Ni(NO_3)_2$ should lead to lower amount of biochar formation as compared with $NiCl_2$ catalyst. The char formation, however, was not studied in detail as it was not the goal of this investigation. Several routes of H_2 formation from a biomass that can be expected include the formation of CO followed by water-gas-shift, direct one step reaction of biomass with high-temperature compressed water, bio-oil formation followed by reaction with water. The bio-oil and its reaction with water generally occurs under pyrolytic conditions. At the end of the reaction, when the reactor contents were cooled down to room temperature, very little amount of the bio-oil was observed in the aqueous solution. Therefore, it is not certain if H_2 was routed via bio-oil formation alone. In the present investigation, two routes appear to be reasonable describing the H_2 generation results. In route one, NO_3^- ions may be accelerating the decomposition of the biomass and Ni^{2+} ions presumably cleaving the C-C bonds and yielding CO and H_2. This may further undergo water-gas–shift reaction and yielding CO_2 and H_2. In route two, after accelerated decomposition of the biomass with NO_3^- ions, Ni^{2+} ions in may be forming water soluble complex with the carbonaceous species derived under hydropyrolysis conditions and yielding H_2 via formation of intermediate compounds.

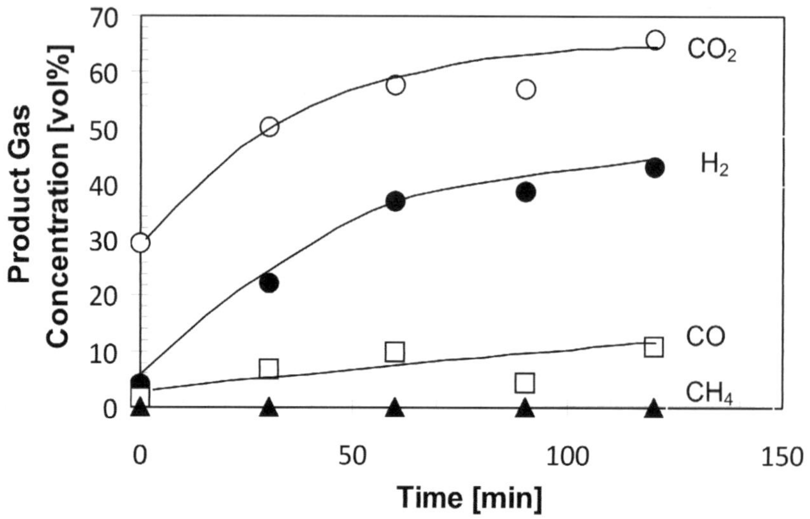

Fig. 3: Product gas profile showing H_2, CH_4, CO and CO_2 gases during subcritical hydropyrolysis of office waste paper at 250°C in presence of 5 wt % $Ni(NO_3)_2$.

The effect of $NiSO_4$ catalyst on the product gas profile during hydrothermal processing of office paper waste at 250°C is shown in Figure 4. Very interesting observations were made. The H_2 formation found to increase almost linearly with the reaction time and reaches to 18vol% in 90 min. This H_2 volume generated is the lowest among all the three Ni-based catalysts investigated here. In 120 min, a total of 24vol% H_2 was observed. The CO_2 formation was almost unchanged from the beginning of the reaction and the volume of CO remains constant after 1 hr of reaction. These trends are significantly different that those observed for the other Ni-catalysts, which suggest different reaction chemistries with $NiSO_4$ catalyst but the biomass liquefaction issue. The office waste paper when treated at different temperatures, a brown colored product slurry containing minimum amount of residue was obtained at 250°C (Figure 5).

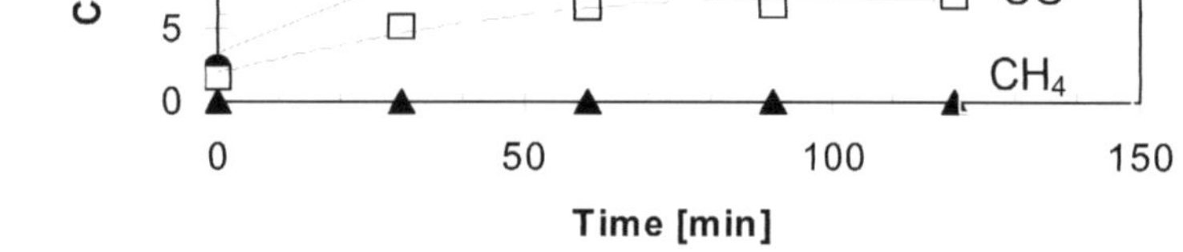

Fig. 4: Product gas profile showing H_2, CH_4, CO and CO_2 gases during subcritical hydropyrolysis of office waste paper at 250°C in presence of 5 wt % $NiSO_4$.

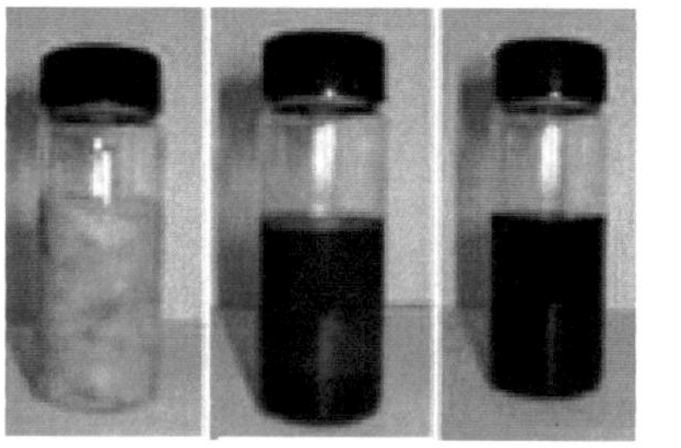

150°C 200°C 250°C

Fig 5: Paper waste slurry processed at different temperatures for 2 hrs.

Composition wise the commonly used office paper contains a major portion of cellulose, about 50 - 95wt%. Therefore, it is obvious that for the H_2 profiles presented in Figure. 2 and Figure. 4, a significant portion must of H_2 have been routed from the cellulose present in office paper waste. To verify this aspect, α-cellulose was subjected to the hydrothermal processing at 250°C in presence of $NiSO_4$ catalyst. The H_2 profile obtained over the period of 2 h reaction time is presented in Figure 6, which indicates about 26 vol% of H_2 generation in 90 min, which is higher by about 8% than that of the H_2 volume observed in the case of office paper waste. It is interesting to note that at the end of 2 h, the volume of H_2 generated for both office paper waste and cellulose were similar indicating that in the case of office paper waste, the H_2 must have been routed from cellulose alone. Small deviation in H_2 volume generated from cellulose and office paper waste may be due to the possible interference or slightly different chemistries associated with the presence of impurities (lignin, colors etc) in the office paper waste.

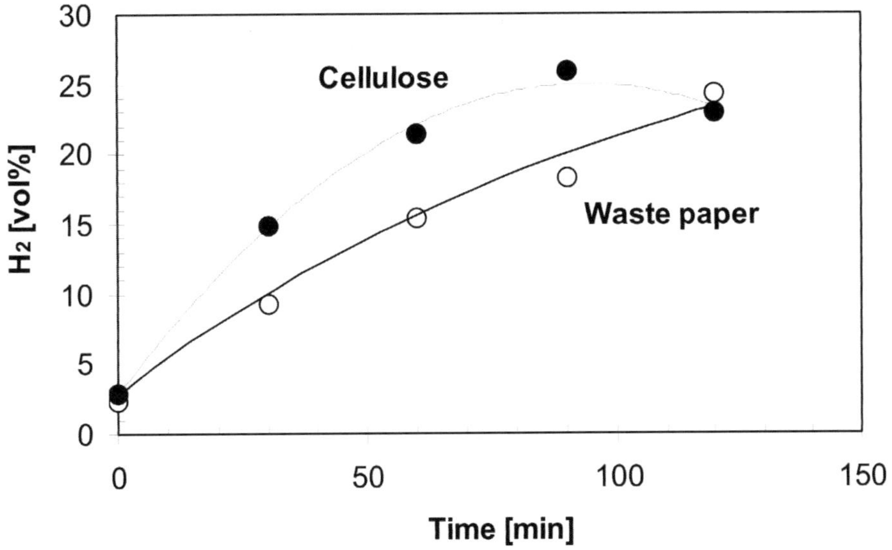

Fig. 6: Comparison of H_2 production from cellulose and waste paper using $NiSO_4$ catalyst at 250°C.

Thus it is possible to achieve significant level of H_2 from the biomass by using homogeneous Ni-catalyst. Such approach does not produce any methane, which is advantageous. From the product gas, the H_2 can be separated using a membrane process and used in the fuel cell for energy generation. The aqueous phase containing Ni-catalyst can be reused again for subsequent biomass processing. Alternatively Ni^{2+} ions can be separated from the processed aqueous stream using ion-exchange resin and the processed water can bedisposed off after recovering the bio-oil and residue. The main advantage of the process investigated here is the minimal amount of char or residue formation.

Conclusion

The subcritical hydrothermal processing of office paper waste and cellulose was successfully performed in high-temperature and high-pressure PARR reactor at 250°C in the presence of homogeneous Ni-catalysts that include $NiCl_2$, $Ni(NO_3)_2$ and $NiSO_4$. Their product gas profiles indicate H_2 rich gas containing CO_2 and CO and CH_4 gases. The concentration of CH_4 was found to be negligible. Almost complete liquefaction of office paper waste was achieved at 250°C with the presence of bio-oil and residue in a very small amount. The maximum H_2 of 43 vol% was produced from office paper waste using $Ni(NO_3)_2$ catalyst whereas $NiSO_4$ generated only 18 vol% after 90 min of reaction at 250°C. In addition to office paper waste, subcritical hydrothermal processing of cellulose was performed at 250°C in the presence of $NiSO_4$. As commonly used office paper contains a major portion of cellulose (approx 50-95%), subcritical hydrothermal processing of cellulose was performed at 250°C in presence of $NiSO_4$ catalyst to determine the extent of H_2 generation. It was observed at the end of 2 hrs that the volume of H_2 generated from both office paper waste and cellulose were very similar indicating that in the case of office paper waste, the H_2 must have been routed from the cellulose.

Acknowledgement

The authors wish to acknowledge the financial supports by the Chemical & Biological Engineering Ph.D. support grant and Center for Bioprocessing Research and Development.

References

1. Li, L., Zhang, H. and Zhuang, X. (2005), “Pyrolysis of waste paper: characterization and composition of pyrolysis oil,” *Energy Sources,* 27, pp. 867 – 873.

2. Torri, N., Okai, A., Shibuki, K., Aida, T. M., Watanabe, M., Ishihara, M., Tanaka, H., Sato, Y., Smith, R. L. and Jr. (2010), “Production of D-glucose from pseudo paper sludge with hydrothermal treatment,” *Biomass and Bioenergy*, 34, pp. 844-850.

3. Ohtsu, Y., Yamada, R., Urasaki, H., Misawa, T., Popescu, S. and Fujita, H. (2010) “Development of novel hybrid microwave- heater reactor for paper -based waste treatment,” *J Mater Cycles Waste Mang*,12, pp. 25-29.

4. Ishida, Y., Kumabe, K., Hata, K., Tanifuji, K., Hasegawa, T., Kitagawa, K., Isu, N., Funahashi, Y. And Asai, T. (2009), “Selective H_2 generation from real biomass through hydrothermal reaction at relatively low temperatures,” *Biomass and Bioenergy*, 33, pp. 8-13.

5. Cheng, L., Ye, P. X., He, R. and Liu, S. (2009), “Investigation of rapid conversion of switchgrass in subcritical water,” *Fuel Processing Technology,* 90, pp. 301-311.

6. Yamaguchi, A., Hiyoshi, N., Sato, O., Bando, K. K., and Shirai ,M. (2010), "Gaseous fuel production from nonrecyclable paper wastes by using supported metal catalysts in high-temperature liquid water," *ChemSusChem,* 3, pp. 737 – 741.

7. Turner, J., Scerdrup, G., Mann, M. K., Maness, P. C., Kroposki, B., Ghirardi, M., Evans, R. J. and Blake, D. (2008), " Renewable hydrogen production," International Journal of Energy Research, 32, pp. 379 – 407.

8. Azadi, P., Khodadadi, A A., Mortazavi, Y. and Farnood, R. (2009) "Hydrothermal gasification of glucose using raney nickel and homogeneous organometallic catalysts," *Fuel Processing Technology,* 90, pp. 145-151.

9. Minowa, T. and Inoue , S. (1999), "Hydrogen production from biomass by catalytic gasification in hot compressed water," *Renewable Energy*, 16, pp. 1114-1117.

10. Antal, MJ, Jr., Manarungson, S, Mok, WS-L. (1994) "Hydrogen production by steam reforming glucose in supercritical water," *Adv. Thermochemical. Biomass Conversion.,* pp. 367-77.

11. Williams, P. T. and Horne, P. A. (1994) "The role of metal salts in the pyrolysis of biomass," Renewable Energy, 4, pp. 1 – 13.

(427c) Hydrogen Production From Liquefied Switchgrass in Supercritical Water Over Ru, Ni, and Co Catalysts

Adam J. Byrd, Sandeep Kumar and Ram B. Gupta, Chemical Engineering, Auburn University, Auburn, AL

Switchgrass is a renewable high yield energy crop with the potential to offset our fossil fuel usage and reduce carbon emissions. In this work, switchgrass was liquefied in subcritical water (T= 300°C, P=100 bar) in a batch reactor to give an aqueous solution of oxygenated hydrocarbons containing ~1.5 wt% C designated as biocrude. Biocrude was then subject to reforming in supercritical water to produce hydrogen. A suite of nine catalysts designated as Metal/Support (M=Ru, Ni, Co; S=TiO2, ZrO2, MgAl2O4) was prepared and evaluated in the supercritical water gasification of liquefied biomass (T=600°C, P=250 bar, WHSV=9h-1). Activity of the catalysts was evaluated in terms of hydrogen yield, gasification efficiency, and tendency towards char formation, as measured by the degree of closure of a carbon balance around the process. Spent catalysts were characterized by the BET method and XRD to evaluate the stability of the support materials in the SCW environment. The highest gasification efficiency was obtained over Ru/ZrO2, while the highest hydrogen yield was seen with Ni/ZrO2.

See more of this Session: Renewable Hydrogen Production I
See more of this Group/Topical: Topical 8: Hydrogen Production and Storage

Modeling of Biohydrogen Production Using BioWin

Hisham Hafez[a*], Elsayed Elbeshbishy[a], M. Hesham El. Naggar[a], George Nakhla[a,b]

[a]Department of Civil Engineering, The University of Western Ontario, London, Ontario, N6A 5B9, Canada

[b]Department of Chemical and Biochemical Engineering, The University of Western Ontario, London, Ontario, N6A 5B9, Canada

Abstract:

Steady state operational data from the patent-pending integrated biohydrogen reactor clarifier system (IBRCS) during anaerobic treatment of glucose-based synthetic wastewater at HRT of 8 hrs and SRT ranging from 26 to 50 hrs and organic loading rates of 6.5 to 206 gCOD/L-d were used to calibrate and verify a process model of the system developed using BioWin. The model accurately predicted biomass concentrations in both the bioreactor and the clarifier supernatant with average percentage errors (APE) of 4.6% and 10%, respectively. Hydrogen production rates and hydrogen yields predicted by the model were in close agreement with the observed experimental results as reflected by an APE of less than 4%, while the hydrogen content was well correlated with an APE 10%. The successful modeling culminated in the accurate prediction of soluble metabolites i.e. volatile fatty acids in the reactor with an APE of 14%. The calibrated model confirmed the advantages of decoupling of the solids retention time (SRT) from the hydraulic retention time (HRT) in biohydrogen production, with the average hydrogen yield decreasing from 3.0 mol H_2 / mol glucose to 0.8 mol H_2 / mol glucose upon elimination of the clarifier.

Keywords: biological hydrogen production, steady state modeling, dynamic modeling, IBRCS

Introduction

Biological hydrogen production from renewable sources (biomass, water, and organic wastes) [1] has the potential to meet the growing demand for energy. It offers a feasible means for sustainable supply of H_2 with low pollution and high efficiency, thereby considered a promising eco-friendly energy source [2]. Comparing the production rates of H_2 by various biohydrogen systems and the

* Corresponding author Tel.: +1 519-630-8683; fax: +1 519-661-3942
E-mail address: hhafez@uwo.ca

associated operational complexity, suggest that dark-fermentation systems offer an excellent potential for practical applications. Although fermentative biohydrogen technologies are still in their infancy, the sustained efforts for developing more efficient technologies will eventually lead to economically feasible technologies. An important step to achieve these feasible solutions includes the development of a process model in order to perform process sensitivity analysis to facilitate the scale-up the process.

Hafez et al. [3] developed a novel integrated biohydrogen reactor clarifier system (IBRCS) [4] (see Figure 1). The system is comprised of a continuously stirred reactor (CSTR) for biological hydrogen production, followed by an uncovered gravity settler for decoupling of solids retention time (SRT) from hydraulic retention time (HRT). The system was able to maintain a maximum hydrogen yield and production rate of 3.1 mol H_2/mol glucose and 191 L/d, respectively. The performance of the IBRCS underscores its potential as an efficient technology that is economically feasible for biological hydrogen production from organic wastes. However, a process model is necessary for better understanding the process and optimization of the system parameters.

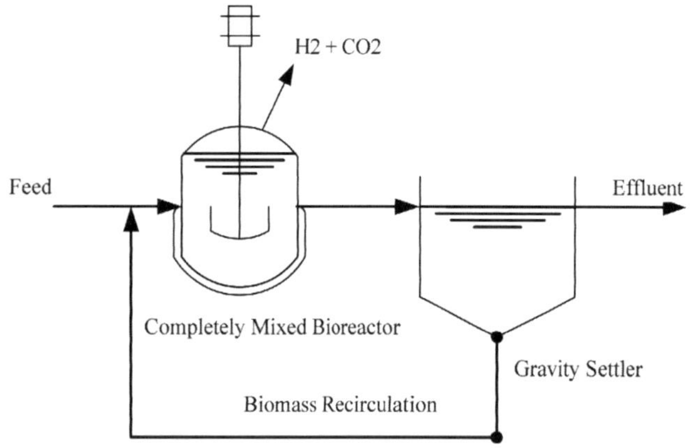

IBRCS [4]

Figure 1. Experimental Setup for the integrated biohydrogen reactor clarifier system

Most studies on biohydrogen production modeling were performed considering batch systems using the Gompertz and Monod equations [5, 6]. The Gompertz equation is an empirical formula where three model parameters (i.e. lag time, H_2 production potential, and H_2 production rate) are adjusted to fit the Gompertz equation to experimental hydrogen evolution data. Even though this curve fitting approach yields high correlation coefficients between the observed and fitted hydrogen evolution data, the three model parameters determined by curve-fitting are restricted to specific experimental conditions and cannot be used in a predictive mode [5]. Due to its empirical nature, the utility of the Gompertz equation is severely limited as it cannot account for relevant process variables such as substrate concentrations, temperature, pH, substrate-types, etc. In some studies, the Gompertz equation has been modified to accommodate typical kinetics of substrate degradation, biomass growth, and hydrogen production [7, 8, 9]. Some biohydrogen studies have utilized the conventional kinetic expressions such as Monod's equation [6, 7, 8, 10, 11]. However, rigorous and multiple simulations, followed by a series of validations may be required to establish the generality of such equations and the associated parameters. Furthermore, comprehensiveness of such models is accomplished only when they can be readily integrated with other complex bioprocesses i.e. hydrolysis, acidogenesis, and H_2 production from complex and particulate organic substrates [5].

A comprehensive literature review demonstrated that no models are readily available that can accurately predict continuous biohydrogen production. On the other hand, most of the available anaerobic digestion models were derived for anaerobic digestion of municipal wastewater biosolids, which employ CSTR and are not capable of modeling a system that decouples the SRT from the HRT (a key feature of the IBRCS). One such model is incorporated in the software BioWin (EnviroSim Associates LTD., Flamborough, Ontario, Canada), which is widely used for modeling wastewater treatment plants. However, it was never used for modeling biohydrogen production systems. Thus, the objectives of this paper are: exploring the use of BioWin model for simulation of biohydrogen production in IBRCS using data from an experimental study [3], evaluating the beneficial effects of decoupling SRT from HRT by comparing the performance of the IBRCS with the conventional CSTR

Materials and Methods

This section provides a brief description of the experimental setup, procedures and operational used in collecting the experimental data for the IBRCS that is later used in the numerical models. Additional information can be found in Hafez et al. [3].

Systems set up and operations

Two lab-scale IBRCSs were considered in the study, each comprising of a continuously stirred reactor (CSTR) for biological hydrogen production (5 L working volume), followed by an uncovered gravity settler (volume 8 L) i.e. open to the atmosphere for the decoupling of solids retention time (SRT) from the hydraulic retention time (HRT). Both systems were operated at 37° C for 220 days (Figure 1), at six different organic loading rates (OLR) ranging from 6.5 to 206 gCOD/L-d. Details of the operational conditions for the six runs are listed in Table 1. It is noteworthy that the systems were run at steady state conditions for at least 20 turnovers of the mean SRT, with the shortest run lasting for 45 days and the longest run for 75 days, excluding the first week of start up. In order to enrich hydrogen producing bacteria, the seed sludges were heat treated at 70° C for 30 minutes prior to startup. Following the completion of each run and the attainment of steady-state conditions, the systems were cleaned and inoculated with pre-treated sludges. OLR-1 and 2 were run simultaneously, followed by OLR-3 and 4, and lastly OLR-5 and 6. The systems were monitored for total chemical oxygen demand (TCOD), soluble COD, volatile fatty acids (VFA), ethanol, lactate, glucose, volatile suspended solids (VSS), total suspended solids (TSS) and biogas composition including hydrogen, methane and nitrogen. The quantity of produced biogas was recorded daily using a wet tip gas meter (Rebel Wet-tip Gas Meter Company, Nashville, TN, USA). Details of analytical methods are reported elsewhere [3].

Table 1. Operational conditions

	Glucose (g/L)	HRT (h)	SRT (h)	OLR (gCOD/L-d)	Final pH
OLR -1	2	8	50 ± 5	6.5	5.5
OLR -2	8	8	45 ± 4	25.7	5.5
OLR -3	16	8	45 ± 6	51.4	5.5
OLR -4	32	8	42 ± 6	103	5.5
OLR -5	48	8	27 ± 3	154	5.5
OLR -6	64	8	26 ± 2	206	5.5

Note. Values represent average ± standard deviation

Inocula and media compositions

Anaerobically-digested sludge from the St.Marys wastewater treatment plant (St.Marys, Ontario, Canada) was used as the seed. The two systems operated in parallel at the same time under two different OLRs for a total of six OLRs (three consecutive runs). The systems were seeded with 5 liters of sludge and started up in a continuous mode with the feed containing glucose at different concentrations as presented in Table 1. The same startup technique was repeated for the three runs. It must be emphasized that there was no sludge wastage from the clarifier throughout the operation, and the values of SRTs presented in Table 1 represent the average ± standard deviation (SD) during steady state operation. It is noteworthy that the reactors operation was consistent over time and accordingly, the average SRT with SD of less than 10% of the mean SRT is representative of the overall SRT during the run.

As expected, the clarifier effluent VSS concentrations were substantially lower than the reactor VSS concentrations and remained unchanged during steady-state operation. The feed contained sufficient inorganics (mg/L): $NaHCO_3$, 2000-16000; $CaCl_2$, 140; $MgCl_2.6H_2O$, 160; NH_4HCO_3, 600; $MgSO_4.7H_2O$, 160; urea, 500-2000; Na_2CO_3, 124-300; $KHCO_3$, 156; K_2HPO_4, 15-20; trace mineral solution, 500; H_3PO_4, 250-1500. Despite the lack of real-time pH control, buffering capacity was sufficient to maintain a final pH of 5.5 in all runs.

Model Formulation

The anaerobic degradation processes in the BioWin model are based on the "four population" model concept (heterotrophs, acetogens, acetoclastic methanogenesis and hydrogenotrophic methanogenesis). To simulate the inhibition of methanogens due to heat pre-treatment of the seed sludge, the methanogens growth was switched off in all modeling runs. A conceptual schematic of the biohydrogen production model is shown in Figure 2. The following points pertaining to the fate of organics in the system describe the key stages:

- Influent biomass undergoes anaerobic decay in the digester. The process rates and stoichiometry are the same as for anaerobic decay in the activated sludge process. The products of decay include unbiodegradable organic, nitrogen, and phosphorus components.

- Hydrolysis of particulate matter is mediated by the (non-polyP) heterotrophs. Particulate matter may be present in the influent, or may consist of products from biomass decay. The products of hydrolysis are phosphate (PO_4-P), soluble organic nitrogen, and readily biodegradable COD.
- Non-polyP heterotrophs ferment the complex readily biodegradable COD to acetic acid, propionic acid hydrogen, and carbon dioxide. There are two model processes for this reaction step. One is for low dissolved hydrogen concentrations while the other is for high dissolved hydrogen concentrations. The stoichiometry of each of these processes is to be calibrated to achieve the appropriate product mix. The calibration can be done by trial and error to achieve the best match between modeled and measured data.
- Propionic acid is converted to acetic acid by acetogenic bacteria. This process also produces hydrogen and is switched off at high levels of propionate. The propionate inhibition constant is 10000 mgCOD/L.

The influent characteristics of the synthetic wastewater containing glucose used in the experimental study was simulated in the model using the influent specifier associated with BioWin model.

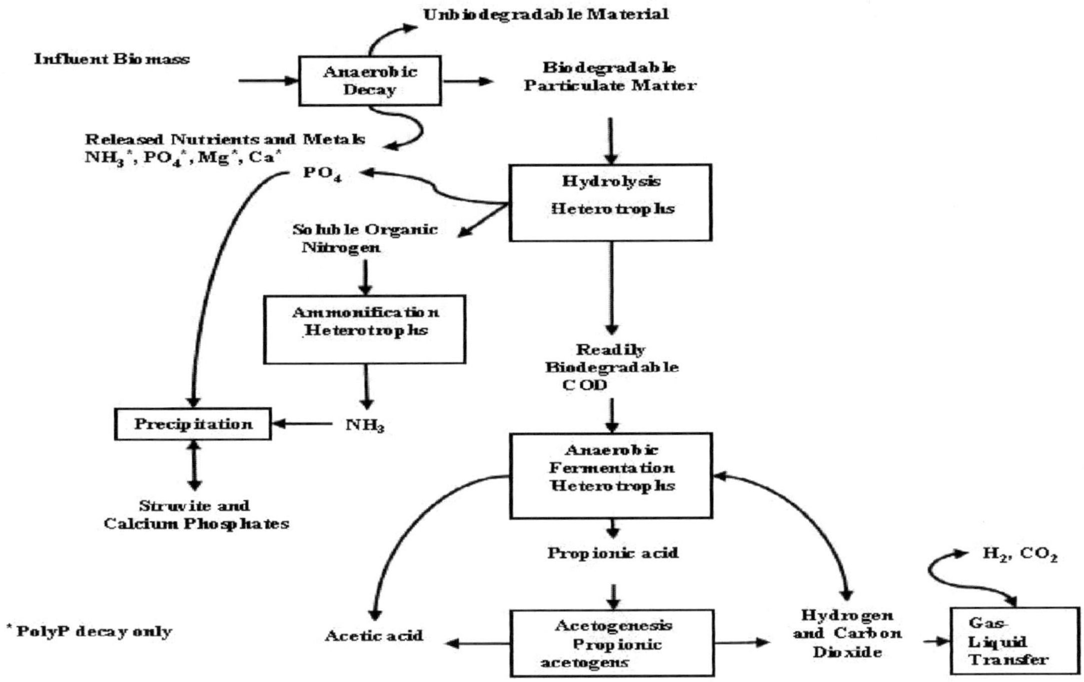

Figure 2. Conceptual schematic for the anaerobic degradation model in BioWin (Adapted from BioWin manual).

RESULTS AND DISCUSSION

Steady State Modeling

Table 2 summarizes the main experimental results under six different organic loading rates, while complete experimental results are reported elsewhere [3]. The model was first calibrated using two sets of experimental data, namely OLR-1 and OLR-2. The calibration involved 180 runs where in tweaking the stoichiometry of the process by trial and error as mentioned earlier was completed to achieve the best fit of the experimental data. It was then validated using the remaining four sets of experimental data (i.e. OLR-3 to OLR-6). Figures 3, 4 and 5 show the correlation between various predicted and measured parameters for the 6 steady state OLRs. The main parameters considered in the comparison were the biomass concentration in the hydrogen reactor measured as volatile suspended solids ($VSS_{reactor}$), biomass concentration in the clarifier supernatant (VSS_{eff}), solids retention time (SRT), main soluble metabolites (i.e. acetate, butyrate vs. propionate), volatile fatty acids (VFAs), soluble chemical oxygen demand in the effluent ($SCOD_{eff}$), hydrogen production rate, hydrogen production yield, and hydrogen content in the biogas.

Table 4. Summary of experimental results

Measured parameters	OLR -1	OLR -2	OLR -3	OLR -4	OLR -5	OLR -6
VSS reactor (mg/L)	1489 ± 116	4190 ± 308	8915 ± 972	15703 ± 926	18472 ± 1404	17038 ± 883
VSS out (mg/L)	247 ± 46	744 ± 50	1578 ± 141	3073 ± 397	5565 ± 581	5240 ± 372
SCOD out (mg/L)	1492 ± 79	6023 ± 194	11922 ± 1230	21267 ± 1627	40960 ± 1624	59091 ± 1358
VFAs (mgCOD/L)	1491 ± 87	5924 ± 257	10344 ± 1114	17976 ± 1444	19232 ± 1156	20582 ± 1686
Hydrogen Gas (%)	71 ± 0.9	73 ± 2.7	65 ± 3.3	67 ± 2.7	43 ± 2.8	39 ± 1.5
Hydrogen Gas (L/d)	12 ± 1.3	48.1 ± 4.7	97 ± 5	179 ± 12	67 ± 5	60 ± 3.4
Yield (mol/mol)	2.8 ± 0.3	2.8 ± 0.3	2.9 ± 0.1	2.8 ± 0.3	1.2 ± 0.1	1.1 ± 0.1

Note. Values represent average ± standard deviation

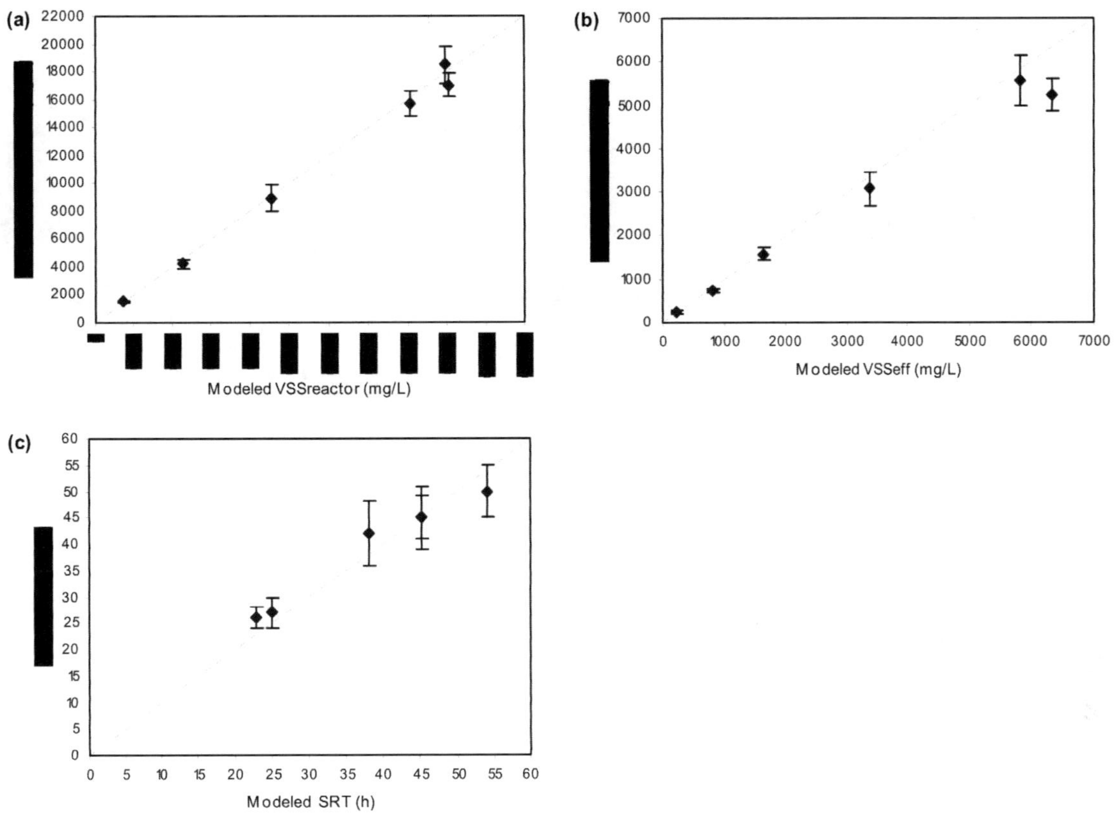

Figure 3. Correlation between modeled and measured parameters for: a) Reactor VSS, b) Effluent VSS, c) SRT (Note: error bars represent standard deviations for experimental results)

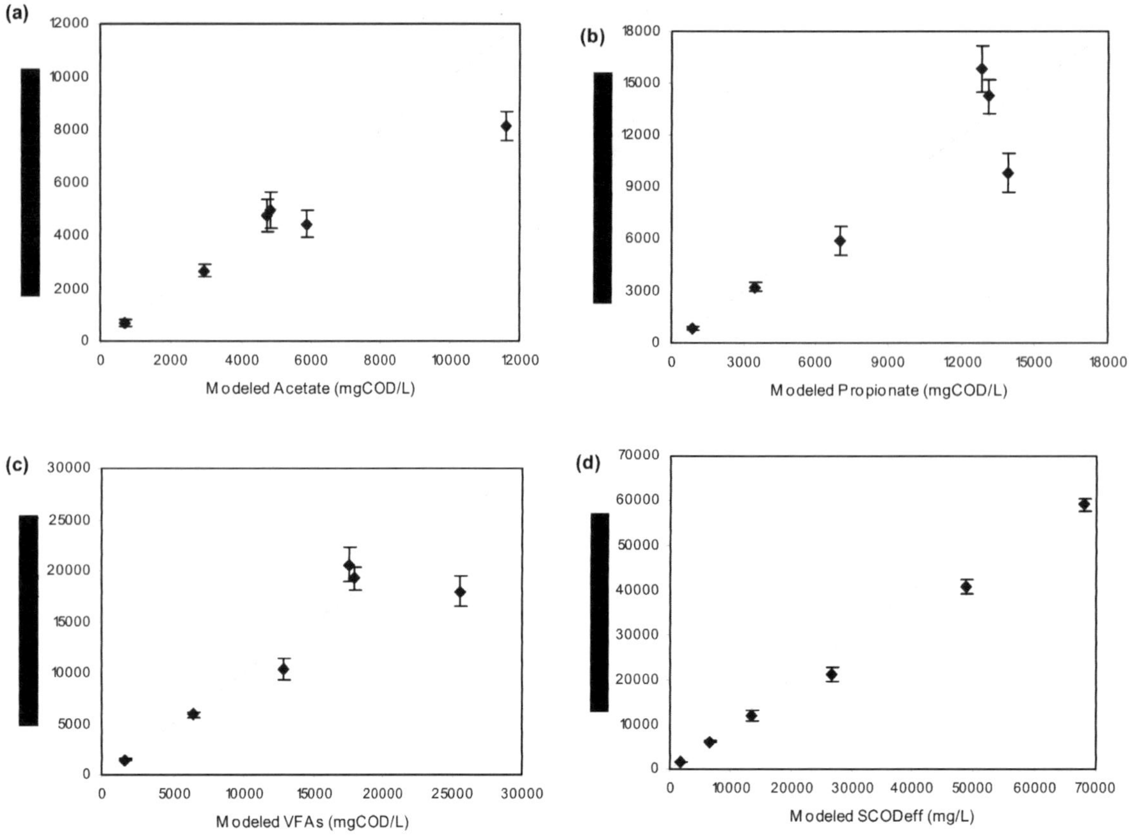

Figure 4. Correlation between modeled and measured parameters for: a) Acetate, b) Propionate vs. Butyrate, c) VFAs, d) Effluent SCOD (Note: error bars represent standard deviations for experimental results)

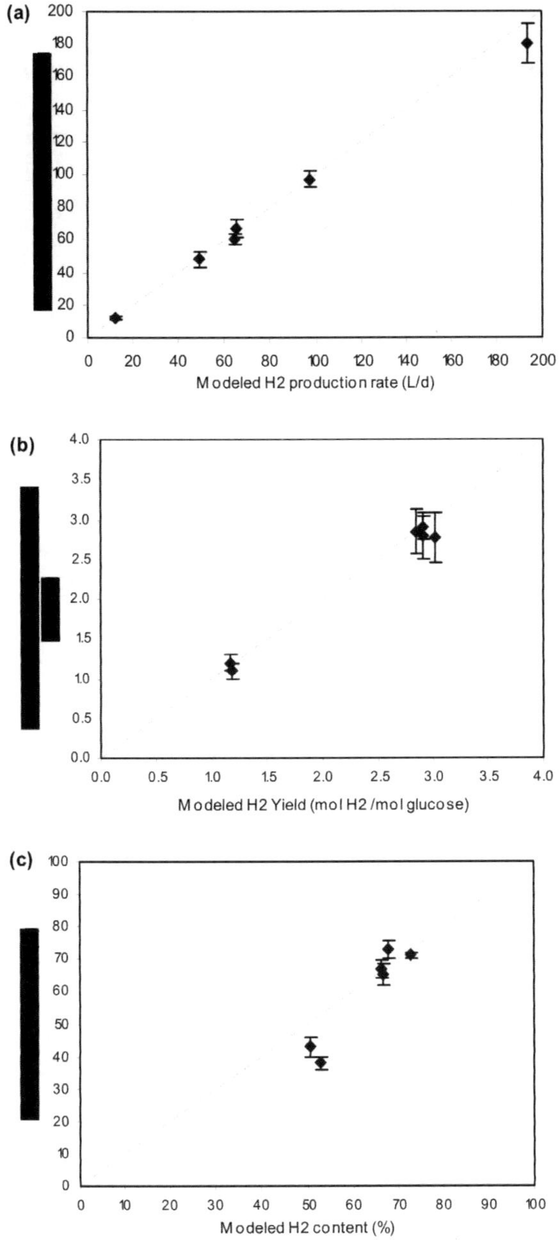

Figure 5. Correlation between modeled and measured parameters for: a) H_2 production rate, b) H_2 yield, c) H_2 content (Note: error bars represent standard deviations for experimental results)

Biomass concentration in hydrogen reactors is a key parameter for the stability and hydrogen production through the process as it directly affects the food-to-microorganisms (F/M) ratio [3]. In the literature, biohydrogen system failures were frequently attributed to marked decrease in biomass content in the hydrogen reactor due to severe cell washout [12, 13]. Figures (3a and 3b) show the correlation between predicted and measured biomass concentrations in both the hydrogen reactor and clarifier effluent for the six experimental runs. As apparent from Figure 3a, the model accurately predicted reactor biomass concentrations throughout the observed range of 1400 to 18000 mgVSS/L at OLRs in the range of 6.5 to 206 gCOD/L-d, with an average percentage error (APE) of 4.6%. The APE is defined as the summation of the absolute difference between the experimental and predicted values divided by the experimental values, averaged over the number of data points. Similarly, the model predictions closely matched the observed biomass concentrations in the clarifier supernatant increase in the range of 200 mgVSS/L to around 6300 mgVSS/L over the same aforementioned range of OLR with an APE of 10% (see Figure 3b). The successful model prediction of the biomass concentrations in both the hydrogen reactor and clarifier effluent was a result of the accurate modeling of the clarifier. As depicted in Figure 3c, the model predicted the solids retention time well for the six experimental runs with an APE of 6.5%.

During the conversion of glucose to hydrogen a mixture of soluble metabolites (i.e. volatile fatty acids) are produced. The two main pathways for glucose conversion are the acetate and butyrate pathways can be given by:

$$C_6H_{12}O_6 + 2H_2O \rightarrow 2CH_3COOH + 2CO_2 + 4H_2 \quad (1)$$

$$C_6H_{12}O_6 \rightarrow CH_3CH_2CH_2COOH + 2CO_2 + 2H_2 \quad (2)$$

Considering Eqs. 1 and 2, the stoichiometric hydrogen yield would be 4 and 2 mol H_2 / mol glucose, respectively. Although BioWin uses an alternative pathway to the butyrate pathway (i.e. glucose → propionate → acetate + hydrogen) (see Figure 2), the predicted and measured acetate and butyrate concentrations agreed well for all experimental runs as can be noted from Figures (4a and 4b), with APEs of 12% and 14.8%, respectively. These results were confirmed by the correlations shown in Figures (4c and 4d) for the total VFAs and the final SCOD with APE of 14% and 13%, respectively.

As apparent from Figure 5, the model accurately predicted both hydrogen production rate and hydrogen yield with a relatively low APE of 4%, while an error of less than 10% was observed when predicting the hydrogen content in the biogas. It is noteworthy that the reactor final pH from both experimental runs and the model were 5.5. The model confirmed the linear increase in hydrogen production rate with the increase of the OLR up to 103 gCOD/L-d. On the other hand, the model predicted an almost constant hydrogen yield of 2.9-3.0 mol H_2/ mol glucose during the same range of OLRs. The steady state model solutions for OLR-5 and OLR-6 have revealed a decrease in hydrogen production rate and hydrogen yield to approximately 65 L/d and 1.2 mol H_2/ mol glucose, respectively.

The high hydrogen yields achieved in the experimental study and corroborated by the model predictions were confirmed by the high biomass concentrations maintained in the hydrogen reactor as well as the operational F/M ratios that fell in the recommended range of 4.4 to 6.4 gCOD/gVSS-d reported in the literature. The authors' earlier work [3] using the experimental data reported here as well as additional data obtained from two CSTRs operated at OLRs of 25.7 and 42.8 gCOD/L-d and HRTs of 8 h and 12 h [13], demonstrated an inverse relationship between the biomass and hydrogen yields (i.e. low hydrogen yields were associated with high biomass yields). On the other hand, the stoichiometric yield of 4 and 2 mol H_2 / mol glucose from Eqs. (1 and 2) are calculated neglecting any biomass yield. Thus, considering the biomass yield would eventually decrease due to energy (COD) utilization for biomass synthesis. Considering COD mass balances, Eqs. 1 and 2 can be modified to calculate the practical ranges of hydrogen yields from glucose conversion to a mixture of acetate and butyrate (neglecting all other factors that might hinder the efficiency), i.e.

$$H_2 \text{ Yield (mol } H_2 \text{ / mol glucose)} = 4\,(1 - Y_{\text{biomass}}) \quad (3)$$

$$H_2 \text{ Yield (mol } H_2 \text{ / mol glucose)} = 2\,(1 - Y_{\text{biomass}}) \quad (4)$$

where Y_{biomass} is the biomass observed yield in gCOD/gCOD.

Assuming a typical biomass yield of 0.15 gCOD/gCOD$_{\text{converted}}$ reported for hydrogen producers [25] and substituting in Eqs. 3 and 4, the practical hydrogen yield should be in the range of 1.7 – 3.4 mol H_2 / mol glucose, while using the widely reported biomass yield of 0.4 gCOD/gCOD$_{\text{converted}}$ in CSTRs [3, 13], the practical hydrogen yield ranges from 1.2 to 2.4 mol H_2 / mol glucose. Equations 3 and 4 clearly justify that in fact an inverse relationship between the biomass and hydrogen yields does exist. Furthermore, based on the values of biomass and hydrogen yields from both experimental and predicted results, the IBRCS achieved a maximum efficiency of 88% (a hydrogen yield of 3.0 mol H_2 / mol glucose compared to a maximum practical H_2 yield of 3.4 mol H_2 / mol glucose).

Evaluation of Decoupling of SRT from HRT

After successful model calibration and verification, the clarifier was removed from the IBRCS to assess the performance of CSTR in comparison to the IBRCS at 4 different glucose concentrations (2 g/L, 8 g/L, 16 g/L and 32 g/L) corresponding to the optimal OLRs 1-4 (6.5 to 103 gCOD/L-d). The performance of the CSTR was strongly affected due to biomass washout. Figure 6 compares the IBRCS and CSTR in terms of reactor VSS, hydrogen yield, hydrogen production rate, and total VFAs. The steady state bioreactor VSS concentration in the CSTR dropped drastically to 120, 500 , 1240 and 1970 mgVSS/L for the OLRs, as compared to 1400, 4550, 9000 and 16100 mgVSS/L in the IBRCS. The extremely high F/M ratios of greater than 40 in the CSTR resulted in a low hydrogen yields ranging from 0.6-1.2 mol H_2 / mol glucose. The maximum hydrogen production rate in the CSTR was approximately 53 L/d compared to a maximum of 194 L/d in the IBRCS and they were both achieved at an OLR of 103 gCOD/L-d. The total volatile fatty acids produced in the IBRCS at the 4 different OLRs were at least 3 folds higher than those produced in the CSTR. Thus, the BioWin model emphatically established the superior performance of the novel patent-pending IBRCS relative to the CSTR.

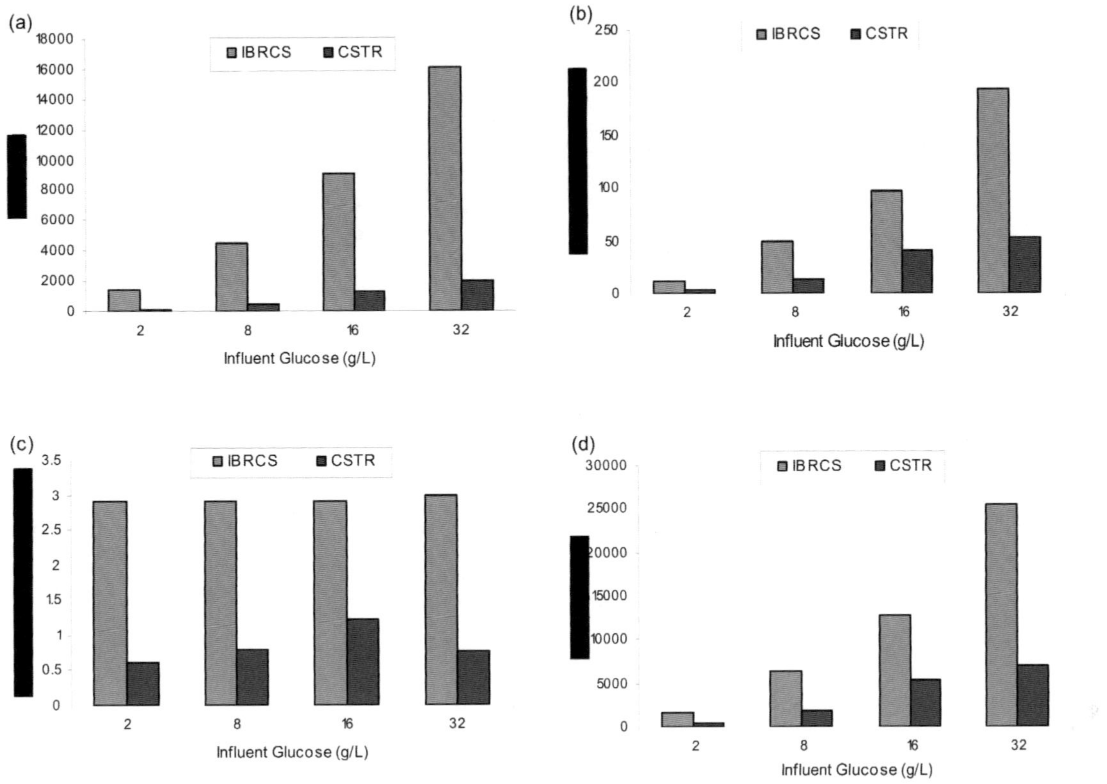

Figure 6. Comparison between the IBRCS and CSTR for: a) Reactor VSS, b) H_2 production rate, c) H_2 yield, d) VFAs

Impact of HRT reduction

To study the impact of HRT reduction on the performance of the IBRCS, three steady state simulation runs at HRT of 4 h, 2 h and 1 h utilizing 32 g/L of glucose were performed in comparison with the steady state data from OLR-4 at 8 h HRT. As depicted in Figure 7, hydrogen production rate decreased by 3%, 13% and 28% with the decrease in HRT from 8 h to 4 h, 2 h and 1 h, respectively. Even though hydrogen production rates seem to be slightly affected by the reduction in HRT, the hydrogen yields were drastically impacted. The hydrogen yield decreased drastically from 3 mol H_2 / mol glucose at an HRT of 8 h to 0.3 mol H_2 / mol glucose at HRT of 1 h. It is noteworthy that the F/M ratio increased from 6.4 gCOD/gVSS-d at HRT of 8 h to 12.8, 26 and 55 gCOD/gVSS-d at HRT of 4 h, 2 h and 1 h, respectively. The increase in F/M ratio was a result of a decrease in the solids retention time due to the increase of solids loading rate to the clarifier which was not sized up accordingly. Thus, it is postulated that sustained higher reactor biomass concentrations with proper clarification would have inevitably improved both hydrogen production rate and hydrogen yield.

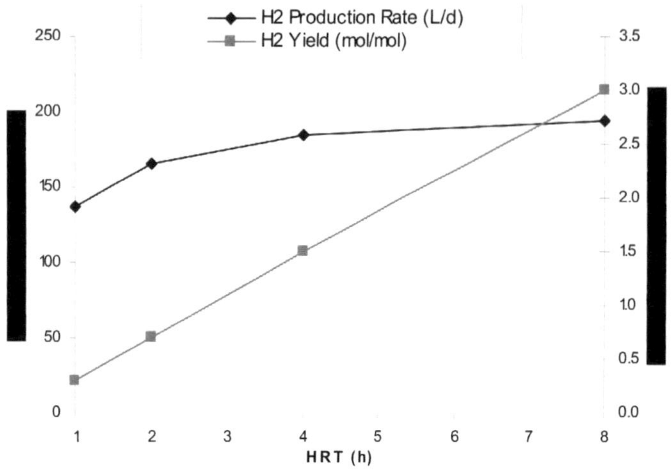

Figure 7. Relationship between hydrogen production rate and hydrogen yield vs. HRT

Summary and Conclusions

Steady state of biohydrogen production in the IBRCS using BioWin was evaluated in this study. The following conclusions can be drawn:

- The model accurately predicted biomass concentrations in both the bioreactor and the clarifier supernatant with average percentage errors (APE) of 4.6% and 10%, respectively. Hydrogen production rates and hydrogen yields predicted by the model were in close agreement with the observed experimental results as reflected by an APE of less than 4%, while the hydrogen content was well correlated with an APE 10%.
- Despite using a different pathway for biohydrogen production, the predicted and measured acetate and butyrate concentrations for the six experimental runs agreed well with APEs of 12% and 14.8%, respectively. These results were validated by the accurate prediction of total volatile fatty acids in the reactor with an APE of 14%.
- The calibrated model confirmed the advantages of decoupling of the solids retention time (SRT) from the hydraulic retention time (HRT) in biohydrogen production, with the average hydrogen yield decreasing from 3.0 mol H_2 / mol glucose to 0.8 mol H_2 / mol glucose upon elimination of the clarifier.

- Steady state simulation for the IBRCS at HRT of 4 h, 2 h and 1 h utilizing 32 g/L of glucose showed that the hydrogen yield decreased drastically from 3.0 mol H_2 / mol glucose at HRT of 8 h to 1.5, 0.7 and 0.3 mol H_2 / mol glucose for HRT of 4 h, 2 h and 1 h, respectively, due to washout as a result of clarifier limitation.

References

1. Cheong, D.Y., and C.L. Hansen. Bacterial stress enrichment enhances anaerobic hydrogen production in cattle manure sludge. Appl Microbiol Biotechnol. 2006; 72: 635–643.
2. Wu, K. J., and J. S. Chang. Batch and continuous fermentative production of hydrogen with anaerobic sludge entrapped in a composite polymeric matrix. Proc Biochem. 2007; 42:279–284.
3. Hafez, H., G. Nakhla, H. El Naggar, E. Elbeshbishy, and B. Baghchehsaraee.. Effect of Organic Loading on a Novel Hydrogen Bioreactor. Int J Hydrogen Energy. 2010; 35: 81-92.
4. Hafez H, Nakhla G, El Naggar H. Integrated system for hydrogen and methane production during landfill leachate treatment. Pending patent, US Patent Application US 61/202,137, 2009.
5. Gadhamshetty V, Arudchelvam Y, Nirmalakhandan N, Johnson DC. Modeling dark fermentation for biohydrogen production: ADM1-based model vs. Gompertz model. Int J of Hydrogen Energy. 2010; 35:479-490.
6. Jianlong W, Wei W. Kinetic models for fermentative hydrogen production: a review. Int J Hydrogen Energy. 2009; 34(8):3313–23.
7. Chen X, Sun Y, Xiu Z, Li X, Zhang D.. Stoichiometric analysis of biological hydrogen production by fermentative bacteria. Int J Hydrogen Energy. 2006; 31(4): 539–549.
8. Chen WH, Chen SY, Kumar KS, Sung S. Kinetic study of biological hydrogen production by anaerobic fermentation. Int. J. Hydrogen Energy. 2006; 31(15):2170–2178.
9. Mu Y, Yu HQ, Wang G. A kinetic approach to anaerobic hydrogen-producing process. Water Res. 2007; 41 (5): 1152–1160.

10. Zheng XJ, Yu HQ. Inhibitory effects of butyrate on biological hydrogen production with mixed anaerobic cultures. J Environ Manage. 2005; 74 (1): 65–70.

11. Lin PY, Whang LM, Wu YR, Ren WJ, Hsiao CJ, Li SL, Chang JS. Biological hydrogen production of the genus Clostridium: metabolic study and mathematical model simulation. Int J Hydrogen Energy. 2007; 32 (12): 1728–1735.

12. Whang LM, Hsiao CJ, Cheng SS. A dual-substrate steady-state model for biological hydrogen production in an anaerobic hydrogen fermentation process. Biotechnol Bioeng. 2006; 95 (3): 492–500.

13. Hafez H, Baghchehsaraee B, Nakhla G, Karamanev D, Margaritis A, El Naggar M.H. Comparative Assessment of Decoupling of Biomass and Hydraulic Retention Times in Hydrogen Production Bioreactors. Int J Hydrogen Energy. 2009;34:7603-7611.

(427e) Nanocomposite Ni-ZrO_2 Catalysts for Ethanol Steam Reforming: Effect of Support Size On the Catalytic Behavior

Xinbin Ma, Shuirong Li, Maoshuai Li, Shengping Wang, Jun Wang and Yongli Sun, Key Laboratory for Green Chemical Technology of Ministry of Education, School of Chemical Engineering & Technology, Tianjin University, Tianjin, China

With heightened concerns for the clean air and clean environment, hydrogen energy derived from renewable sources in conjunction with fuel cell has recently been under investigation actively. Producing hydrogen from steam reforming of ethanol is considered as a preferred method. Ni-based catalyst is highly recommended in this process while the oxide supports play an important role on its selectivity and stability. Researches in the nanomaterials recently have revealed that catalytic nanoarchitecture with comparably sized metal and oxide domains can retain high activity because of an increase the interfacial contact area and hindering agglomeration of the active metal. It opens a new route in the development of highly active and stable catalysts for the steam reforming process. Here we present primary study on the impacts of zirconia particle size on the catalytic behavior of nanocomposite Ni-ZrO_2 catalysts in the steam reforming of ethanol. Three nanocomposite Ni-ZrO_2 catalysts with particle sizes of zirconia concentrated at 50 nm, 25 nm and 15 nm were carefully prepared by precipitation followed by impregnation method and nominated here as Ni-ZrO_2-50, Ni-ZrO_2-25 and Ni-ZrO_2-15 respectively. The average nickel particle sizes in the first two catalysts were about 62 nm and 23 nm after reduction. To better understand the size effect of zirconia support, the Ni-ZrO_2-15 catalyst underwent specially treated to enlarge the nickel particles to 80 nm in the reduced samples. The impact of weight flow rate on the catalytic activity was investigated at 400 ≡C, ethanol/H_2O = 1:3 for all the three catalysts. The catalysts showed complete conversion of ethanol at W/F higher than 2.0×10-4 (gcat h)/ml but dropped in an order of Ni-ZrO_2-50, Ni-ZrO_2-15 and Ni-ZrO_2-25 catalysts in the reverse direction, indicating different activities of the catalysts. Analysis of the selectivity towards different carbonous compounds in the reactions operating at condition range that incomplete conversion of ethanol was achieved reveals clue in the reaction pathway of different catalysts, in which a main route of ethanol undergoing dehydrogenation to produce acetaldehyde followed by decomposition of acetaldehyde into CO and CH_4 was found for all the three catalysts. Besides, trace of acetone and ethylene were also detected in the product stream of the Ni-ZrO_2-50 and Ni-ZrO_2-15 catalysts while little was found for the Ni-ZrO_2-25 catalyst under the reaction condition range studied here. In ethanol steam reforming, acetaldehyde could also be oxidized into acetic acid and further transform into acetone by ketonization reaction, which could lead to unfavorable coke deposition on the surface of the support. Zirconia is reported to have high surface oxygen mobility especially in tetragonal phase when comparing with monoclinic phase. In the Ni-ZrO_2-15 catalyst with a reduced zirconia particle size under the size limitation, predominant tetragonal phase along with trace of monoclinic phase of zirconia were detected in the XRD analysis while only a single monoclinic phase of zirconia was detected in the reduced Ni-ZrO_2-50 and Ni-ZrO_2-25 catalysts. The increase of oxidation activity due to the reduction of zirconia particle size has led to a higher selectivity towards acetone formation in the ethanol steam reforming for Ni-ZrO_2-15 catalyst. In spite of the improvement in oxidation activity of the catalyst, the reduction in particle size of zirconia would not have significant positive impact on the catalytic activity unless the particle size of the nickel metal is also reduced into a limited range. Although the zirconia and the nickel particles are in a comparable range, 50 nm seems to be too large for the Ni-ZrO_2-50 catalyst to perform a good activity in the steam reforming of ethanol. Reducing the particle of zirconia while keeping the nickel metal particle size in a range of larger than 50 nm, as in the situation of the Ni-ZrO_2-15 catalyst, the improvement in the catalytic activity could only be in a limited range since insufficient C-C bond cleavage activity is supplied by the nickel metal. When both of the particle sizes of the zirconia and the nickel metal are reduced into a suitable range as in the situation of the Ni-ZrO_2-25 catalyst, the increase of metal support interaction as revealed in the H_2-TPR profiles lead to a significant improvement in both the activity and selectivity of the catalyst.

See more of this Session: Renewable Hydrogen Production I
See more of this Group/Topical: Topical 8: Hydrogen Production and Storage

(427f) Hydrogen Production From Ammonia Using a Micro-Reformer System Integrated with a Heat-Recirculating Micro-Combustor

Jung Ho Kim[1], Sang Ik Lee[1], Jae Moon Joo[1] and Oh-chae Kwon[2], (1)School of Mechanical Engineering, Sungkyunkwan University, Suwon, Gyeonggi-do, South Korea, (2)School of Mechanical Engineering, Sungkyunkwan University, Suwon, South Korea

In order to evaluate the potential of reforming ammonia as a carbon-free fuel in production of hydrogen, an annulus-type micro reforming system integrated with a micro-combustor is studied experimentally. The micro-combustor is a simple cylinder with an annular-type shield to apply for a heat-recirculation concept and an expanded exhaust outlet to control ignition, satisfying the primary requirements for designing a micro-combustor as a heat source, i.e., stable burning in a small confinement and uniform distribution of temperature along a wall. The micro-combustor is surrounded with a micro-reformer for converting ammonia gas to hydrogen. The annulus-type micro reforming system is designed to produce 1-10 W (based on lower heating value, LHV) of hydrogen using various catalysts (Ru, Ni/SiO2/Al2O3 or Ir), which satisfies the primary requirements for designing an effective micro reforming system, i.e., maximum heat transfer through the micro-combustor wall and enhanced hydrogen production. The feed rate of ammonia, the micro-combustor inlet velocity of fuel-air mixtures and the catalyst materials substantially affect the performance of the designed micro reforming system. Under optimized design and operating conditions, the micro reforming system using Ru produces 5.4 W (based on LHV) of hydrogen with a conversion rate of 98% and an overall system efficiency of 13.7%. Thus, the present configuration can be applied to practical micro reforming systems, supporting the potential of using ammonia as a clean fuel.

See more of this Session: Renewable Hydrogen Production I
See more of this Group/Topical: Topical 8: Hydrogen Production and Storage

Effect of Ultrasonication on Pulp Waste Solubilization and Enhancement of Hydrogen Production in Batch Study.

Elsayed Elbeshbishy, Hisham Hafez, George Nakhla*

Dept. of Civil and Environmental Engineering, University of Western Ontario, London, Canada, London, Ontario, Canada, N6A 5B9

Abstract

Batch anaerobic studies were conducted to study the effect of ultrasonication as a pretreatment method for pulp waste prior to anaerobic digestion for hydrogen production in batch reactors. The ultrasonication pretreatment was conducted by sonicating a 100 mL of pulp waste for different sonication times varied from 0.5 min to 30 min. The ultimate hydrogen production increased with increasing the sonication time. The highest ultimate hydrogen production was achieved at sonication time of 30 min with about 88% increase compared with the unsonicated food waste. The maximum hydrogen production rate at sonication time of 30 min was marginally (about 145%) higher than that the unsonicated food waste.

Keywords: Ultrasonication, Food waste, Hydrogen, Pretreatment.

2. Introduction

Research on alternative energy sources has gained renewed interest, due to global awareness of accumulated carbon dioxide in the atmosphere as a potential cause of climate change [1]. Combustion of H_2 produces no greenhouse gases, and has a high-energy yield of

* Corresponding author. Tel.: +1 519 661 2111x85470; fax: +1 519 850 2991.
E-mail addresses: gnakhla@fes.engga.uwo.ca, gnakhla@eng.uwo.ca (G. Nakhla).

122 kJ/g, which is 2.75-fold greater than that of hydrocarbon fuels [2]. However, the rate of biological H_2 production is low and the technology needs further development [3]. Thus, the improvement of bio-hydrogen producing efficiency is an urgent requirement for its industrialization [4]. The major criteria for substrate selection are the availability, cost, carbohydrate content, and biodegradability [5]. Commercially produced food products, such as corn and sugar, are not yet economical for hydrogen production. Alternatively, wastewaters with organic waste such as food processing and animal waste have great potential as substrate sources [6]. Treating these waste streams to protect public health and the environment while producing a clean energy source makes biological hydrogen production an attractive alternative to fossil fuels [5]. Food industry wastes constitute a major fraction of the municipal solid wastes. High carbohydrate content in form of simple sugars, starch and cellulose makes the food wastes a potential feedstock for biological hydrogen production. Therefore, food industry wastes meet all the abovementioned criteria, which can make them ideal candidates for hydrogen production via microbial processes [7].

Ultrasonication has been increasingly used recently as a pre-treatment method for anaerobic digestion. Although ultrasonication is widely used as pretreatment method to solubilise the substrate and then enhance the methane production, there is a lack of substrate pretreatment for biohydrogen production. Thus, the primary objective of this work was to study the effect of ultrasonication on food waste solubilization and therefore enhancement of hydrogen production.

2. Experimental set-up

The food waste obtained from city of Toronto, Ontario, Canada was used as substrate, the average characteristics of this food waste in (mg/L) were; TCOD: 91902, SCOD: 49878, TS: 65480, VS: 46140, particulate carbohydrate: 26489, soluble carbohydrate: 19976, particulate protein: 6254, and soluble protein: 8711. The VFAs was 1990 mg COD/L. Anaerobic sludge was collected from the primary anaerobic digester at St Mary's wastewater treatment plant (St Mary's, Ontario) and used as seed sludge. The TSS and VSS concentrations of the sludge were 11,200 and 7,600 mg/L, respectively. The lab scale ultrasonic probe used in this study was supplied by Sonic and Materials, Newtown, USA (model VC-500, 500 W, and 20 kHz). 100 mL of food waste was sonicated for different sonication times (0.5, 2.5, 5, 10, 20, and 30 min) corresponding to different specific energy inputs (0.35, 1.2, 3, 5.5, 15, and 23 kJ/g TS), with sonication pulses set to 2 seconds on and 2 seconds off. To control the temperature rise of the sludge, a cooling water bath was used, and the sludge temperature during the experiments did not exceed 30 ^{0}C.

Batch anaerobic studies were conducted using the sonicated food waste as substrates and the anaerobic digester sludge as seed to assess hydrogen production rates. The batch experiments were conducted in triplicates in a series of serum bottles (liquid volume of 200 ml). The volumes of substrate (food waste) and seed (anaerobic digester sludge) calculated based on food to microorganisms (F/M) ratio of 4 on TCOD to VSS basis. For the blank, the substrate volume was replaced by distilled water. To each bottle, 50 mL of food waste and 150 mL of seed were added. The initial pH value of the mixture in each bottle was adjusted to 5.0 using 1N NaOH and HCl. The headspace was flushed with oxygen-free nitrogen gas for a period of 3 minutes and capped tightly with rubber stoppers. The bottles were then placed in a swirling-action shaker (MaxQ 4000, Incubated and Refrigerated Shaker, Thermo Scientific, CA) operating at 180 rpm and maintained at a temperature of 37 ^{0}C. Control bottles were prepared using the food waste without sonication.

3. Results and discussion

3.1. Ultrasonication and food waste solubilization

Ultrasonication can disrupt part of the particulate components and convert it from particulate to soluble components [21]. Due to ultrasonication pretreatment, the TS, VS, particulate carbohydrate, and particulate protein were decreased, while the SCOD, soluble carbohydrate, soluble protein, and VFAs were increased. As shown in Table 1, the SCOD increased sharply with increasing the sonication time up to 10 min with 18.6% increase compared with the unsonicated one. After 10 min sonication, a little increase in SCOD was observed (only 3.5 % after 30 min sonication time, from 18.6% to 22.1%). The same behaviour for soluble carbohydrate and soluble protein was observed, the soluble carbohydrate increased by 29.1% and 30.3% at 10 at 30 min sonication times, respectively. While the soluble protein did not have that high increase as the SCOD and the carbohydrate, it increased by only 11% and 13.6% at 10 and 30 min sonication times, respectively. VFAs increased by about 38% after 5 min sonication and after that there was no significant increase in the VFAs. On the other hand, all the particulate components decreased with increasing the sonication time up to 20 min, after which, there was no significant decrease in all of those components. The TS, VS, particulate carbohydrate, and particulate protein decreased by 12%, 14%, 19%, and 12% after 20 min sonication time, respectively. Based on the abovementioned results, it is evidence that the ultrasonication pretreatment can enhance the releasing of carbohydrate and protein into the liquid phase which therefore enhance the hydrogen production.

3.2. Hydrogen production

Batch experiments were conducted to study the effect of ultrasonication at different sonication times of food waste on the hydrogen production. The unsonicated food waste was examined as well. The batch experiment showed that the biogas production contained only

Table 1. Percentages increase and decrease in different components.

sonication time	SE (kJ/kg TS)	% Decrease				% Increase			
		TS	VS	Particulate Carbohydrate	Particulate Protein	SCOD	Soluble Carbohydrate	Soluble Protein	VFA
No sonication	0	0	0	0	0	0	0	0	0
0.5 min	350	1.7	1.2	10.7	0.8	0.2	7.6	2.4	8.0
2.5 min	1200	4.6	5.4	19.6	2.0	4.3	12.1	7.6	12.3
5 min	3000	6.4	7.2	14.4	5.3	6.9	23.8	8.5	38.2
10 min	5500	9.5	10.4	17.2	8.4	18.6	29.1	11.1	37.4
20 min	15000	12.0	14.1	18.8	11.9	20.4	29.7	11.1	40.2
30 min	23000	11.5	13.8	21.9	12.5	22.1	30.3	13.6	40.3

hydrogen and carbon dioxide, without detection of methane. The cumulative hydrogen production form the unsonicated food waste and the sonicated one at different sonication times are shown in Figure 1 (all the experiments were conducted in triplicate and the error bars are not shown in this figure as error was less than 12%).

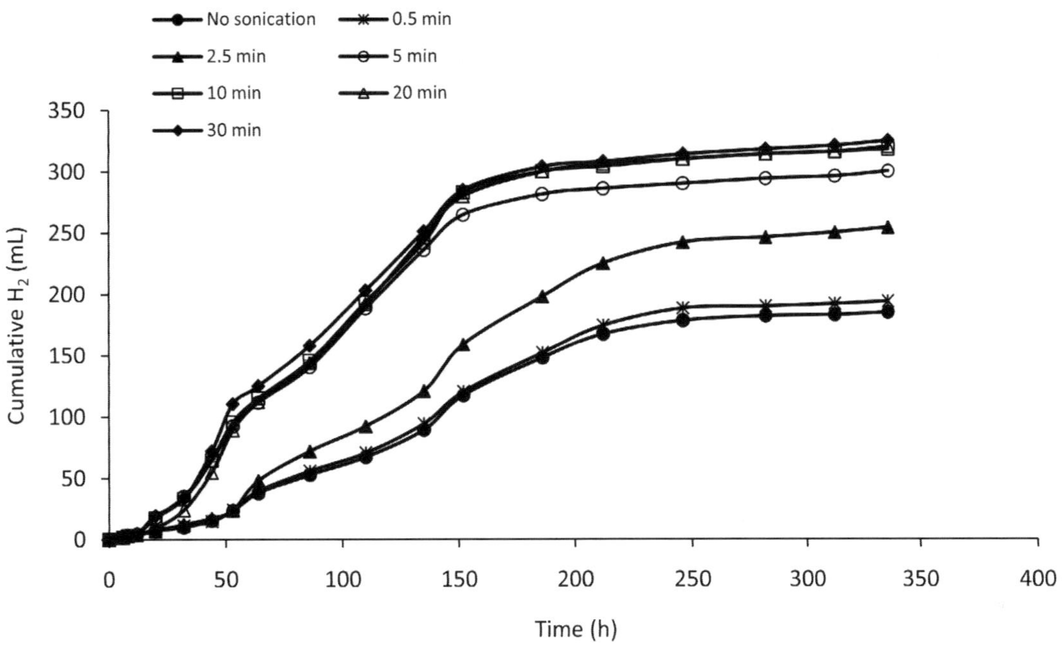

Figure 1. Cumulative hydrogen production for sonicated and unsonicated food waste.

As shown in Figure 1, the ultimate hydrogen production for the sonicated food waste was higher than that of the unsonicated food waste (185 mL). As apparent from the figure, by increasing the sonication time, the ultimate hydrogen production increases. The highest ultimate hydrogen production was achieved of 325 mL at sonication time of 30 min, which is about 77% increase. After 0.5 min sonication, there was no significant increase in hydrogen production, only about 5% (from 185 to 194 mL), while for the 2.5 min sonication time the ultimate hydrogen production increased from 185 to 254 mL. For the 5, 10, and 20 min sonication times, ultimate hydrogen productions were 300, 318, and 320 mL, respectively, which are corresponding to 62%, 73%, and 74% increase compared to the unsonicated food waste.

Figure 2 illustrates the hydrogen yield for the unsonicated food waste and the sonicated one at different sonication times. The lowest hydrogen yield of 80 mL/g VS_{added} was observed for the unsonicated food waste, while the highest hydrogen yield of 141 mL/g VS_{added} was achieved at 30 min sonication time. The hydrogen yield of 110, 130, 138, and 139 mL/g VS_{added} was observed for the 2.5, 5, 10, and 20 min sonication times, respectively.

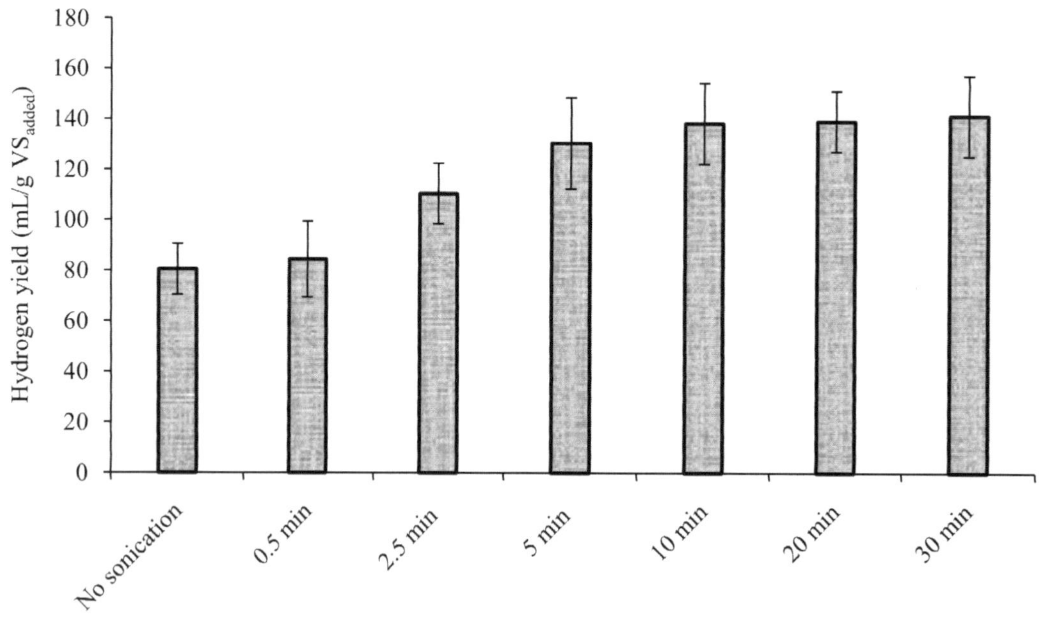

Figure 2. Hydrogen yield for sonicated and unsonicated food waste.

The maximum increase in hydrogen yield was about 77% at 30 min sonication time compared with the unsonicated one. The hydrogen yield after 0.5 min sonication was about the same as the unsonicated food waste (84 mL/g VS_{added}). Based on the abovementioned results, it is evident that with increasing the sonication time, the hydrogen yield increases up to 10 min after which there was no significant effect of the ultrasonication.

Figure 3 shows the final VFAs after the fermentation for the unsonicated and sonicated food wastes at different sonication times. For the unsonicated food waste and the sonicated food waste at 0.5 and 2.5 min, the VFAs ranged from 1835 to 2158 mg COD/L. The VFAs of the sonicated food waste at sonication times of 5, 10, and 20 min were ranged from 2985 to 3065 mg COD/L. The highest value of the final VFAs after fermentation of 3125 mg COD/L was achieved at sonication time of 30 min, which is corresponding to 70% increase compared to the unsonicated food waste.

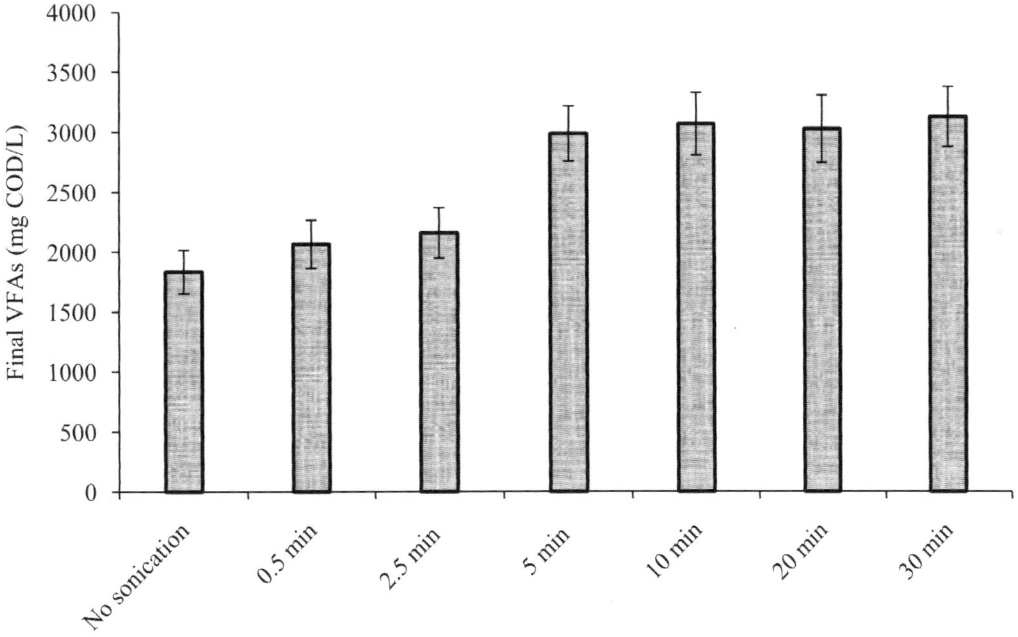

Figure 3. Final VFAs after fermentation for sonicated and unsonicated food waste.

4. Conclusion

The outcome of this study emphatically revealed the positive effect of sonication on food waste solubilization and biological hydrogen production. Based on the findings of this study, the following conclusions can be drawn:

- The ultrasonication pretreatment enhanced the releasing of carbohydrate and protein into the liquid phase which therefore enhanced the hydrogen production.
- The lowest hydrogen yield of 80 mL/g VS_{added} was observed for the unsonicated food waste, while the highest hydrogen yield was 141 mL/g VS_{added}.
- The highest value of the final VFAs after fermentation was achieved at sonication time of 30 min, which is corresponding to 70% increase compared to the unsonicated food waste and there was no significant difference between the acetate to butyrate ratios (HAc/HBu) for the all samples.

Acknowledgements

Authors want to thank Trojan Technologies and Egyptian Ministry of Higher Education for their financial contributions.

References

[1] D. Das, T.N. Veziroglub, Hydrogen production by biological processes: a survey of literature, *Biochem Eng J.* 23 (2005) 221-230.

[2] M.L. Cai, J.X. Liu, Y.S. Wei, Enhanced biohydrogen production from sewage sludge with alkaline pretreatment, *Environ Sci Technol.* 38 (2004) 3195-3202.

[3] D.B. Levin, L. Pitt, M. Love, Biohydrogen production: prospects and limitations to practical application, Int J Hydrogen Energy 29 (2004), 173–85.

[4] Y.P. Guo, S.Q. Fan, Y.T. Fan, C.M. Pan and H.W. Hou, The preparation and application of crude cellulase for cellulose-hydrogen production by anaerobic fermentation, *Int J Hydrogen Energy* 35 (2010) 459–468.

[5] I.K. Kapdan, F. Kargi, Bio-hydrogen production from waste materials, Enzyme Microbial Technol 38 (2006), 569–582.

[6] J. Benemann, Hydrogen biotechnology: progress and prospects, Nature Biotechnology 14 (1996), 1101-1103.

[7] I. Ntaikou, G. Antonopoulou, G. Lyberatos, Biohydrogen Production from Biomass and Wastes via Dark Fermentation: A Review, Waste Biomass Valor 1 (2010), 21–39.

[8] E. Elbeshbishy, S. Aldin, H. Hafez, G. Nakhla, M. Ray, Impact of Ultrasonication of Hog Manure on Anaerobic Digestability, Ultrasonics Sonochemistry (2010), doi:10.1016/j. ultsonch.2010.04.011.

(491b) Modeling and Operation of a Photo-Electrochemical Reactor for H2 Production

Chris Carver[1], Zack Ulissi[1], **Klaus Hellgardt**[2], Chin Kin Ong[1], Stephen Dennison[1] and Geoffrey Kelsall[1], (1)Imperial College London, London, United Kingdom, (2)Chemical Engineering, Imperial College London, London, United Kingdom

Sustainable fuels are required to contribute to the overall energy carrier portfolio of the world and to substitute some of the more polluting energy vectors we depend on today. For some decades now has the conversion of solar energy to usable energy (electricity or chemical fuel) been an important scientific goal. The direct conversion into electricity has been achieved and is now been driven by industry to provide ever cheaper devices for solar to electricity conversion. The production of devices of solar based fuels has been somewhat slower in its deployment, mainly due to the perceived high device costs and the intrinsic complexities involved but also due to the lower conversion that can be currently achieved. Photo-electrochemical systems, which facilitate the production of molecular hydrogen and oxygen gas directly from water and sunlight, have been pursued for the last 40 years, ever since the pioneering work by Fujishima and Honda in 1971. However, the primary focus in this area has been on the development of photo-electrode materials, for high efficiency and low-cost manufacture. The development of practical photo-electrochemical reactors for larger-scale hydrogen production has not yet received significant attention. Photo-electrochemical water splitting is a very demanding process, requiring materials that are capable of both absorbing solar radiation efficiently and remaining stable under the conditions of hydrogen or oxygen evolution. The successful nano-structuring of, especially, lower band-gap materials such as WO3 and Fe2O3, has reinvigorated interest in photo-electrochemical approaches to solar hydrogen production. Fe2O3, produced by a chemical vapor deposition process has been found to produce photocurrents greater than 0.2 A m-2 under AM1.5 sunlight. If this performance can be extended to large area electrodes, fabricated by an inexpensive route, then large-scale photo-electrolyzers based on Fe2O3 photo-anodes become a practical possibility. We have designed and built deposition systems that allow the inexpensive production of 0.01 m2 photo-anodes. These photo-anodes can be accommodated in a newly developed photo-electrochemical reactor (Figure 1), which we aim to study in order to develop a complete understanding of the issues involved in the scale-up of photo-electrochemical reactors for hydrogen production. In this context we have built a multi-physics model to describe the fluid flow as well as the current density distribution within the reactor and thus consider the operational consequences of bubble formation and sheet/electrode conductivity, channel spacing etc. Figure 1: Schematics of a flat plate photo electrochemical reactor for operation and scale-up analysis.

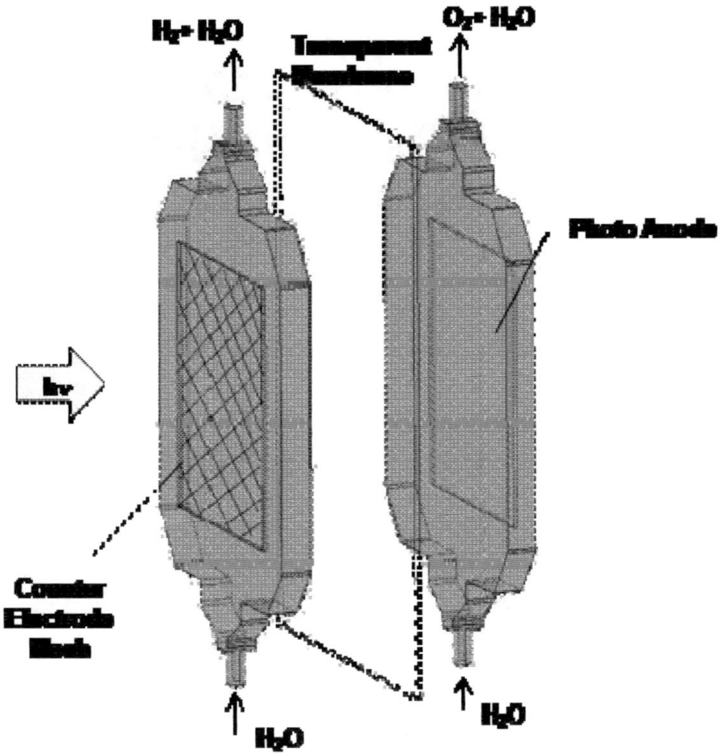

See more of this Session: Renewable Hydrogen Production II
See more of this Group/Topical: Topical 8: Hydrogen Production and Storage

(491c) Non-Thermal Plasma Reforming of Refractory Tars and Oils Generated by Biomass Gasification

Joseph J. Hartvigsen[1], Lyman Frost[2], S. Elangovan[2], Michele Hollist[1] and Piotr Czernichowski[1], (1) Ceramatec, Inc., Salt Lake City, UT, (2)Ceramatec, Inc, Salt Lake City, UT

One of the barriers to commercial implementation of biomass gasification is the treatment refractory tars and oils created in the gasification process. Typical downstream operations such as a combustion turbine, reciprocating engine genset, or biomass to liquids (BTL) catalyst bed all require a tar free gasifier product stream for successful operation. These tars and oils represent lost mass and energy if simply removed from the gasifier product stream. Removal and disposal of tars and oils also produces a hazardous material stream creating both real and regulatory penalties to the gasifier operator.

Ceramatec, as a sub-recipient of a DOE funded research project, will be applying its non-thermal plasma reformer technology to convert refractory tars and oils while still entrained in the gasifier product stream. Preliminary laboratory work indicates that it is possible for the plasma reformer to operate on the gas exiting the gasifier and convert these tars and oils to additional hydrogen and carbon monoxide. This work is being done at Western Research Institute in Laramie, Wyoming under a US Department of Energy award to Emery Energy.

This paper will present an overall process structure of the gasification process and the non-thermal plasma reformer in particular. Operational characteristics and laboratory results related to the reforming of the refractory tars and oils will be presented. The larger pilot sized facility at Laramie will also be described. Data may be available from the larger pilot sized facility in Laramie by the time for the conference and if available will be presented.

See more of this Session: Renewable Hydrogen Production II
See more of this Group/Topical: Topical 8: Hydrogen Production and Storage

(491d) Synthesis of Functional Nanostructured TiO2 Photo Active Membranes for Room Temperature Water Splitting

Alok Vats[1], **Anurag Y. Kawde**[2], Rajesh Shende[1] and Jan A. Puszynski[1], (1)Chemical and Biological Engineering, South Dakota School of Mines and Technology, Rapid City, SD, (2)Chemical and Biological Engineering, South Dakota School of Mines & Technology, Rapid City, SD

Photo assisted splitting of water at room temperature to produce hydrogen on a liquid-semiconductor (L-S) interface holds great technological importance toward hydrogen economy. In order to electrochemically split water into H_2 and O_2, thermodynamic electrochemical potential (E_0) of 1.23 V (Vs NHE) is required, where $E_{reduction}$ is at 0 V(Vs NHE) for H_2 half cell reaction ($2H^+ + 2e^-$ àH_2) and $E_{oxidation}$ is -1.23 V(Vs NHE) for O_2 half cell reaction ($2OH^-$ à $2H^+ + 2e^- + O_2$). The above reactions occurs preferentially on a L-S interface under solar radiation. Various semiconductors such as CdS, InP, SiC, Si, TiO_2, Ta_2O_3, Fe_2O_3 and $SrTiO_3$ have been studied as a potential photo-catalyst for photolysis of water. However, majority of these materials have shown a poor stability to photo oxidative corrosion with TiO_2 being an exception. TiO_2 is a broadband semiconductor with E_g ~ 3.1eV (UV) and a relatively poor electronic conductor. However, its electronic properties can be improved by controlling its shape and chemical composition. This contribution focuses on the synthesis and application of specific TiO_2 based functional nanostructures in a form of a membrane for photo assisted water splitting application. The proposed functional membrane structure has an advantage over the conventional approach i.e. colloidal TiO_2. In the present work, an attempt has been made to take advantage of superior directional electronic transport in TiO_2 nanotubes (nt-TiO_2). Under this approach, an nt-TiO_2/Ti/nt-TiO_2 membrane structure was synthesized using electrochemical anodization in fluorine (F^-) based electrolyte. Further, this nt-TiO_2/Ti/nt-TiO_2 membrane structure was modified by preferential deposition of Pt and RuO_2 co-catalysts on either side of the membrane. The selective placement of Pt and RuO_2 on nt-TiO2/Ti/nt-TiO2 membrane structure could prove to be beneficial to achieve physical separation of H_2 and O_2 in a water photolysis reactor utilizing a single electrode scheme.

See more of this Session: Renewable Hydrogen Production II
See more of this Group/Topical: Topical 8: Hydrogen Production and Storage

(491e) Oxygen Inactivation of An [FeFe] Hydrogenase From *Clostridium Pasteurianum* and Evolution of a Mutant Hydrogenase with Decreased Oxygen Sensitivity

Alyssa S. Bingham and James R. Swartz, Department of Chemical Engineering, Stanford University, Stanford, CA

Hydrogenase enzymes rapidly catalyze the reversible reduction of protons with electrons to produce hydrogen. These enzymes have the potential to be used in photosynthetic hydrogen production or fuel cell applications, where energy can be captured from sunlight, and the reactants and byproducts are water and oxygen. One main obstacle in industrial hydrogenase use is the oxygen sensitivity of the protein. Here we investigate the oxygen inactivation of an [FeFe] hydrogenase from *Clostridium pasteurianum* and describe the evolution of a mutant with decreased oxygen sensitivity.

We show that, in contrast to previous observations, [FeFe] hydrogenase I from *Clostridium pasteurianum* (CpI) can be partially reactivated after oxygen exposure. We observe slow reactivation of CpI over time in a 2% hydrogen environment in the presence of methyl viologen, and rapid reactivation of the hydrogenase in the presence of a strong reducing agent such as dithionite. After reactivation, the hydrogenase activity is partially but not fully restored. This observation suggests that there are two states of inactivation of CpI by oxygen: one reversible and one irreversible. We further show that this reactivation does not occur for the [FeFe] hydrogenase HydA1 from *Chlamydomonas reinhardtii*, which suggests that the accessory [Fe-S] clusters present in CpI but not HydA1 may be responsible for this reversible inactivation.

Additionally, we have used random mutagenesis and a cell-free protein synthesis platform to produce and screen a library of CpI hydrogenase mutants for improved oxygen tolerance, and have identified a mutant that is inactivated more slowly in the presence of oxygen than the wild-type enzyme. The initially isolated mutant had 13 mutations, and three of these mutations contribute to the decreased oxygen sensitivity of the protein. Saturation mutagenesis at one of these sites has resulted in further improvement of the mutant. We show, however, that both the mutant and wild-type enzyme are reactivated to the same extent in the presence of an electron source, and the mutant is therefore not immediately useful for biohydrogen applications. This demonstrates the need to develop a screen that measures hydrogenase activity while the enzyme is working in the presence of oxygen in order to evolve an oxygen-tolerant hydrogenase for biohydrogen applications.

PERFORMANCE EVALUATION OF SULFUR TRIOXIDE DECOMPOSER IN A SMALL-SCALE TEST LOOP AT KAERI

Jong-Ho Kim*, Sung-Deok Hong, Tae-Hoon Lee, Ki-Young Lee, Yong-Wan Kim

Korea Atomic Energy Research Institute, Republic of Korea

** corresponding author : jhkim5@kaeri.re.kr*

1. Introduction

Korea Atomic Energy Research Institute (KAERI) is developing hydrogen production process called thermo-chemical SI(Sulfur Iodine) cycle utilizing the heat from the High Temperature Gas Cooled Reactor (HTGR) with outlet coolant temperature up to 950 °C, which is considered as an efficient reactor for the hydrogen production[1]. One of the key components in the SI cycle is the sulfur trioxide decomposer, which is operating under very high temperature, high pressure, and very corrosive sulfuric acid condition. The small scale test loop (10kW) was installed at KAERI site to confirm the integrity of the key components of SI cycle under the severe operational conditions. The KAERI test loop consists of primary and secondary system as shown in Fig. 1[2]. In this paper, the sulfur trioxide decomposer was simulated using aspen plus chemical process simulator to predict the performance under the test loop operating condition. The simulation flow-sheet and material balance for the sulfur trioxide decomposer were developed in case of a 0.5 kg/min of sulfuric acid flowrate. The parametric study was carried out for sulfur dioxide production in the sulfuric acid decomposer. The result shows that moles of sulfur dioxide produced at the sulfur trioxide decomposer increases as the operating temperature increases. It also shows that the higher the pressure is, the lower the sulfur dioxide production rate is at a given temperature.

2. Description of Test Loop

Figure 1 shows the schematic diagram of the small-scale test loop, which includes the primary and secondary system. The primary loop consists of circulator, sulfur trioxide decomposer, pre-heater, main heater, hot gas duct and nitrogen supply system. The design conditions of the primary loop side are;

- Working fluid : Nitrogen
- Design Temperature : 950 °C
- Design Pressure : 40 atm
- Design Flow : 2.0 kg/min

The secondary loop consists of H2SO4 feed pump, H2SO4 evaporator, H2SO4 decomposer, SO3 decomposer, cooler, and SO2 trap etc. The SO3 and H2O decomposed from H2SO4 is flowing in the secondary loop as a gas and its flow rate is controlled by a mass flow controller. The design conditions of the secondary loop side are ;

- Working fluid : H2SO4
- Design Temperature : 850 °C
- Design Pressure : .50 atm
- Design Flow : 1.0 kg/min

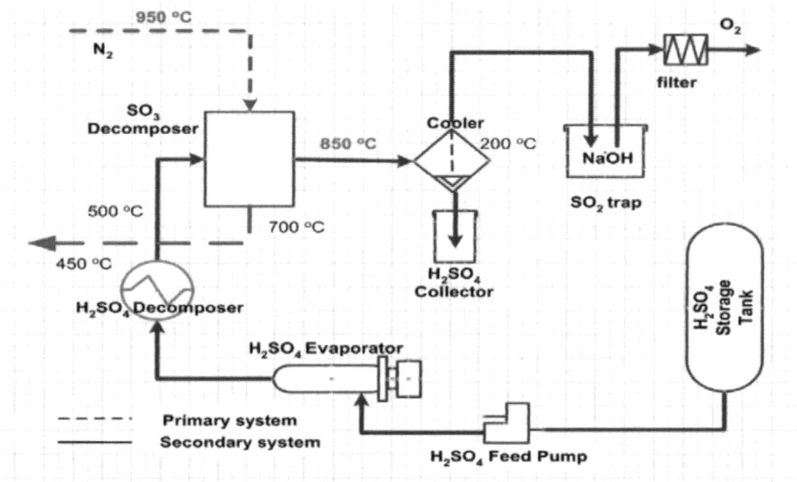

Fig. 1. Schematic Diagram of Small-scale KAERI Test Loop

The liquid sulfuric acid with 98% by weight concentration is injected to the system at 7 atm by a sulfuric acid feed pump. The liquid sulfuric acid is heated to the vaporization temperature at the sulfuric acid evaporator before its decomposition. During the heating process, the liquid sulfuric acid begins to vaporize and the mixture of gas and liquid is formed inside the sulfuric acid evaporator. The sulfuric acid evaporator is designed so that only the sulfuric acid vapor is formed at the exit of the sulfuric acid evaporator. In the sulfuric acid decomposer, decomposition of gaseous sulfuric acid takes place. It proceeds further as the gaseous sulfuric acid stream is heated through the sulfuric acid decomposer. The sulfuric acid decomposer is designed so that most of the sulfuric acid vapor decomposes into sulfur trioxide and water before it exits the sulfuric acid decomposer. The gaseous products from the sulfuric acid decomposer are sent to the sulfur trioxide decomposer. Decomposition of sulfur trioxide into sulfur dioxide and oxygen takes place in sulfur trioxide decomposer. The product gases from sulfur trioxide decomposer are finally cooled to the condensing temperature of the water in the cooler. In the cooling process, the remaining sulfur trioxide reacts with water vapor to form sulfuric acid, which is collected in the sulfuric acid collector.

3. Simulation Approach[3,4]

3.1 Components

The components involved in the simulation are water, sulfuric acid, sulfur dioxide, sulfur trioxide, oxygen, nitrogen. Components H2O and H2SO4 are chosen for the electrolytes, and acidic species are treated as hydronium ion. Hence the hydronium ion(H3O+) , bisulfate ion, and sulfate ion are additionally listed as a corresponding electrolytic ions.

3.2 Chemistry Models

Chemistry models have been used to accurately describe the complex chemical behaviors such as dissociation or complexation in the the evaporation or disintegration phenomena of sulfuric acid at high temperature and high concentration. The typical chemistry models are; 1) H2SO4 + H2O <--> H3O+ + HSO4- , which is important at all concentrations, 2) HSO4- + H2O <--> H3O+ + SO4= , which is important in weak sulfuric acid solutions.

3.3 Reaction Models

Reaction models have been used to describe decomposition phenomena of sulfuric acid gas and sulfur trioxide gas more accurately. The chemical reactions in this simulation include sulfuric acid evaporation, sulfuric acid liquid/gas decomposition, and sulfur trioxide gas reactions. Reaction models for gas reactions are ; 1) SO2 + 0.5 O2 <--> SO3 ,which is kinetic, catalytically enhanced, and 2) SO3 + H2O <--> H2SO4 . Reaction models are more practical, when used together with chemistry models because they leads to a better model convergence.

3.4 Physical Properties

The ElecNRTL model is used to properly generate electrolyte species and reactions. It calculates liquid phase properties from the Electrolyte-NRTL activity coefficient model. Henry's Law is used to calculate gas (SO2, O2, N2) solubility in sulfuric acid. It was assumed in the simulation that all single phase gases or mixture gases of SO2(g), SO3(g), O2(g), H2O(v) were considered ideal, and apparent component approach was used.

3.5 Simulation for the Sulfur Trioxide Decomposer

Sulfuric acid is heated to the vaporization temperature before decomposition. It begins to vaporize during which some of the sulfuric acid decomposes into SO_3 and H_2O. The sulfuric acid decomposer retrieves much of the heat remaining after sulfuric acid decomposition. The decomposition reaction of sulfuric acid proceeds further as the vaporized stream is heated through the sulfuric acid decomposer. Most of the sulfuric acid decomposes into sulfur trioxide and water vapor before it exits the sulfuric acid decomposer. Decomposition of sulfur trioxide into sulfur dioxide and oxygen is performed in sulfur trioxide decomposer. The sulfur trioxide decomposer is modeled as a combination of stoichiometric reactor and a heater in the simulation.

4. Simulation Results

The simulation flowsheet was determined as in Fig.2. The material balance around the sulfur trioxide decomposer was obtained as in Table 1, which shows the temperature, pressure and molar flow rate for each stream.

4.1 Simulating Flowsheet and Material Balance

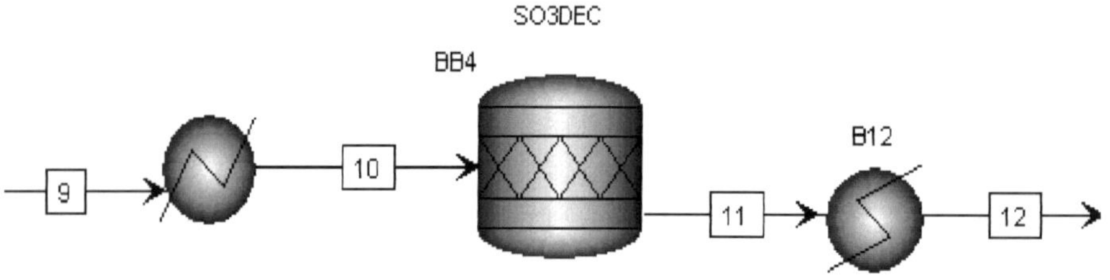

Fig. 2. Simulating Flowsheet for the Sulfur Trioxide Decomposer

Table 1. Material Balance around the Sulfur Trioxide Decomposer

Streams	9	10	11	12
Temperature, C	750	850	850	430
Pressure, atm	6.9973	6.9973	6.9973	6.99729
Mole Flow,kmol/hr				
H2O	0.6599	0.6599	0.6599	0.65994
H2SO4	0.0025	0.0025	0.0025	0.00251
SO3	0.5978	0.5978	0.2286	0.22855
SO2	0	0	0.3693	0.36928
O2	0	0	0.1846	0.18464
Mole Frac				
H2O	0.5236	0.5236	0.4567	0.45673
H2SO4	0.002	0.002	0.0017	0.00174
SO3	0.4744	0.4744	0.1582	0.15818
SO2	0	0	0.2556	0.25557
O2	0	0	0.1278	0.12779
Total Flow,kmol/hr	1.2603	1.2603	1.4449	1.44492

4.2 Parametric Study for Sulfur Trioxide Decomposer

In sulfur trioxide decomposer 1mole of sulfur trioxide decomposes into 1 mole of sulfur dioxide and 0.5 mole of oxygen as in the following reaction.

$SO3 \rightarrow SO2 + 1/2\ O2$

The reaction takes place in a gas phase since the operating temperature and pressure is 850 °C and 7 atm, respectively.The parametric study was carried out for SO2 production in the sulfur trioxide decomposer. In Fig. 3, moles of SO2 produced are represented as a function of temperature in case of 1, 5, 7, 10 atmospheres of SO3 decomposer operating pressures. It is shown that moles of SO2 produced increases with the operating temperature increase and/or with operating pressure decrease.

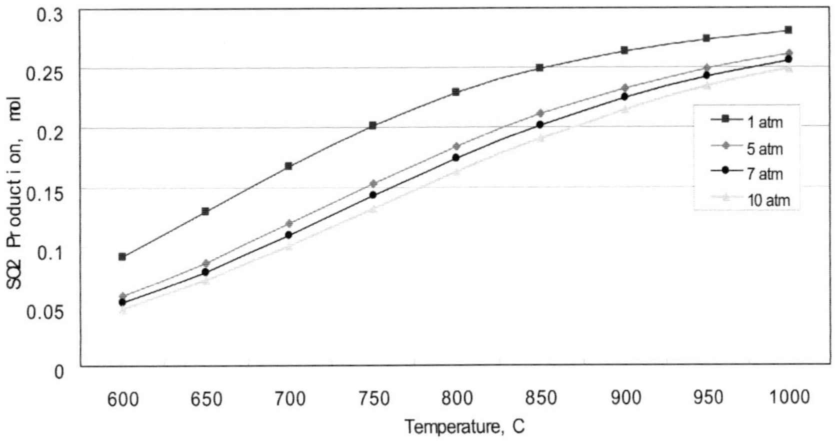

Fig. 3. Moles of SO2 Produced as a Function of Temperature

In Fig. 4, moles of SO2 produced are represented as a function of moles of sulfuric acid that was used in the secondary system.

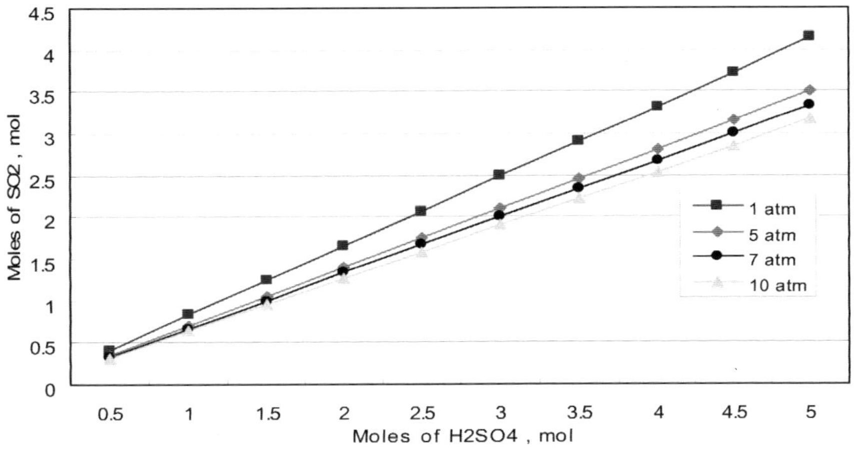

Fig.4. Moles of SO2 Produced as a Function of H2SO4 Content

In Fig. 5, moles of the mixture that exist inside the SO3 decomposer are represented as a function of temperature. It is shown the moles of SO3 decreases with operating temperature, while the moles of SO2 and oxygen are increasing as the decomposition reaction proceeds.

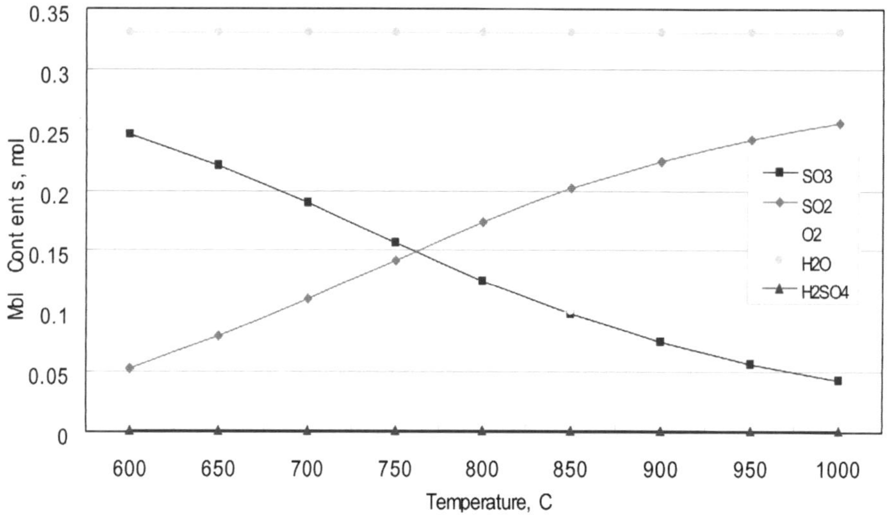

Fig. 5. Mole Contents Change during SO3 Decomposition Reaction

5. Conclusions

The small-scale test loop(10kW) was designed conceptually and installed at KAERI site to test the integrity of the sulfur trioxide decomposer under high T and P operating conditions. The performance of the sulfur trioxide decomposer is predicted by simulation to incorporate the result into the design of the corresponding components in SI cycle coupled with HTGR. In the simulation, material balance around the sulfur trioxide decomposer was determined and simulation flowsheet was obtained. The parametric study was carried out for sulfur trioxide decomposer. It is shown that moles of SO2 produced at the sulfur trioxide decomposer increases as the operating temperature increases. It is also indicated that the production rates of sulfur dioxide is decreasing as the operating pressure becomes higher at a given temperature.

ACKNOWLEDGEMENTS
This work was financially supported by the Korean Ministry of Education, Science and Technology.

References

[1] J. H. Chang et al., A study of a Nuclear Hydrogen Production Demonstration Plant, Nuclear Engineering and Technology, Vol. 39, No. 2, pp. 111-122
[2] S. D. Hong et. al. Development of a Compact Nuclear Hydrogen Coupled Components (CNHCC) Test Loop, American Nuclear Society 2007, Embedded Topical Meeting ST-NH2, Jun 25-30 Boston, USA, 2007
[3] Aspen Plus version 2006, Aspen Technology, 2007
[4] BROWN L.C. etc., High Efficiency Generation of Hydrogen Fuels using Nuclear Power, General Atomics Company, GA-A24285, 2003
[5] Ozturk I.T. etc., An improved process for H2SO4 decomposition step of the sulfur- iodine cycle, Energy Conversion and Management 36 (1),11- 21, 1994

(506a) Hybrid Sulfur Cycle Options for the Next Generation Nuclear Plant

Maximilian B. Gorensek, Computational Sciences Directorate, Savannah River National Laboratory, Aiken, SC and William A. Summers, Energy Security Directorate, Savannah River National Laboratory, Aiken, SC

The hybrid sulfur (HyS) cycle is one of the two high-temperature water-splitting processes under consideration for use with the Next Generation Nuclear Plant (NGNP) demonstration at Idaho National Laboratory. This presentation will describe the reference case HyS flowsheet developed for NGNP and discuss the synergies and challenges of integrating HyS with other potential NGNP applications.

See more of this Session: Advances in Thermochemical Hydrogen Production
See more of this Group/Topical: Topical 8: Hydrogen Production and Storage

Development of Non-Pt Electrocatalysts in HBr Proton Exchange Membrane Electrolyzer for Hydrogen Production

Rui Zhang, John W. Weidner

Center for Electrochemical Engineering, Department of Chemical Engineering, University of South Carolina, Columbia, SC 29208

1. Introduction

Hydrogen could be produced via indirect splitting of water at a lower temperature (800-1000°C) than in the direct thermal dissociation (>2500°C) [1-4], through the promising thermal-chemical reaction of modified hybrid calcium-bromide-based cycle. The whole cycle basically consists of a thermo-chemical reaction to produce O_2 and an electrolytic process to produce H_2. In the electrolytic process, Br_2 is evolved at the anode with the feed of anhydrous HBr to promote mass transfer, while hydrogen evolution reaction occurs at the cathode. The electrode reactions are [1]:

Anode: $$2HBr \rightarrow Br_2 + 2H^+ + 2e^- \qquad E^\circ = 0.58 \text{ V vs. SHE}$$

Cathode: $$2H^+ + 2e^- \rightarrow H_2 \qquad E^\circ = 0\text{V vs. SHE}$$

Pt, as a standard electrode, has been widely investigated for both reactions above [5-10]. However, scarcity and high cost of Pt would impede scaling-up and commercialization of this electrolyzer. What's more, strong adsorption of Br^- on the Pt would significantly decrease its activity towards HER at the cathode side [7-9], when HBr crossover takes place. Therefore, non-Pt catalysts need to be developed for both half reactions to mediate the cost and improve the electrode stability.

In this paper, RuO_2, Vulcan XC 72R carbon black and Niobium doped TiO_2 are prepared onto the anode side of the membrane electrode assembly and tested in the HBr electrolyzer. MoS_2 and IrO_x are prepared by a low temperature method, prepared into a thin film electrode, tested on the rotating disk electrode towards HER.

2. Experimental

HBr PEM Electrolyzer

The HBr PEM electrolyzer uses the same testing system as previous one [10], except the cell hardware. The cell hardware uses the aluminum backing plate of the 10 cm^2 cell purchased from Fuel Cell Technologies, Inc. The flowing fields on both sides were grooved on a composite of graphite and fluoro inert polymer (Diamond F100), with 0.031 inch width and 0.04 inch depth of the groove. A couple of Kynar plates were placed between the aluminum ending plates and flowing fields to prevent the ending plates from corroding by hydrogen bromide. Also copper current collectors are followed between the Kynar plates and backing plates.

MEAs (Membrane electrode assembly) with different electrode materials are applied and usual procedure for assembling the cell was followed. Data points of current and cell potential are measured and collected on HP power supply.

MEA Preparation

MEA (membrane electrode assembly) was prepared by the regular spray and hot-pressing method. The carbon cloth (totally hydrophilic) from ETEK was used as the gas diffusion layer. At the anode, 1.5 mg/cm^2 catalysts were sprayed directly onto the carbon cloth substrate. Nafion 212 (poly [perfluorosulfonic] acid membrane) from Dupont was used as the electrolyte. The nafion membrane was boiled in 1 N H_2SO_4 solution for 90 minutes and rinsed thoroughly with de-ionized water. The commercial 0.5mg/cm^2 Pt/C EDL (ELAT from ETEK) served as the cathode directly. Thus derived sandwich was placed in a Carver hot press (Model # 3851-0) for hot-pressing at 500 tons and 120ºC for 2 minutes. Cool the MEA in the air and save it ready to use.

RDE Experiment

Synthesis of MoS_2 and IrO_2/C

MoS_2 is prepared by the solvent-heat method [11] by reacting sulfur (Alfa Aesar) and $Mo(CO)_6$ (Alfa Aesar, 98%) in p-xylene. In a typical synthesis, 30.6 mg sulfur and 126 mg $Mo(CO)_6$ powders are dissolved in 100 mL p-xylene in a three-neck round bottom flask. Nitrogen is purged to protect the solution from air. Then, this solution is heated to 140 °C and held for three hours. The black precipitate (MoS_2) is then separated from the solution via filtering and dried in the vacuum oven at 90 °C for about 4 hours. In the post-treatment experiment, the material is heated to 450 °C and 800 °C under nitrogen in a horizontal furnace.

IrO2 is prepared from EG Method [12]. Measure $RuCl_3$ or $IrCl_3$ into a crucible and incipient wetness with ethylene glycol (EG). Sonicate the mixture for 30 minutes to obtain a homogeneous mixture. Add carbon and stir with a spatula until fully moistened. Sonicate again for 30 minutes. Dry the mixture at 140°C for 12 hours in vacuum oven. Grind dried solids into a fine powder. Heat sample to calcinate and hold for one hour. Samples were calcinated at 300°C, 350°C, and 400°C.

Preparation of the film electrode

The film electrode was made from depositing proper catalyst ink onto the disk electrode from Pine.The catalyst ink is prepared by mixing 6 mg of solid catalysts with 3 mL Isopropyl Alcohol (IPA, >99%, Sigma-Aldrich) and 3 mL water (MilliQ water, $\rho > 18$ MΩcm). The ink was sonicated for 20 minutes and then pipetted onto the disk electrode to gain the loading of 20 μg solid catalysts. To make the catalyst film uniform, a drop of 5 μL catalyst ink at most was pipetted stepwise and 5 μL of binding solution was finally applied and dried after the desired catalyst loading was achieved. The binding solution was prepared by mixing 5% wt. Nafion in IPA solution (Ion-power, Inc) and additional IPA in the volume ratio of 1:20.

Catalyst characterizations

Electrochemical characterizations are performed by cyclic voltammetry (CV) and steady state polarization in a three-electrode setup from Pine Instrument in the aqueous solution of 1.0 M $HClO_4$ as the electrolyte. The reference electrode is Hg/Hg_2SO_4 (-0.68 V vs. NHE) and a

Pt wire was used as the counter electrode. Before each test, the electrolyte was de-aerated with purging N_2 for 10 minutes. Potential is swept at the rate of 50 mV/s under the control of the Princeton Applied Research 263A potentiostat controlled by Corrware (Scribner Associates Inc.). After holding at the open circuit for 20 minutes, the potential starts from the open circuit potential in the positive direction and swept back and forth in the potential range set. Pseudo-Steady state polarization was conducted by potentiodynamic method at 2 $mV \cdot s^{-1}$ at the same potential window. The current densities were all normalized to the geometric area of the working electrode.

Characterization of bulk structure is done on the powder X-ray Diffraction (XRD, Rigaku miniflex II) and the microscopy of the surface morphology is done on High Resolution Transimission Electron Microscopy (HRTEM, JEOL JEM 2100F)

3. Representative Results and Discussion

The primary performance of the vapor phase HBr PEM electroyzer is shown in figure 1. It is seen that HBr electrolytic cell has higher efficiency and productivity than HCl process. Gaseous phase feeding of HBr could improve mass transfer.

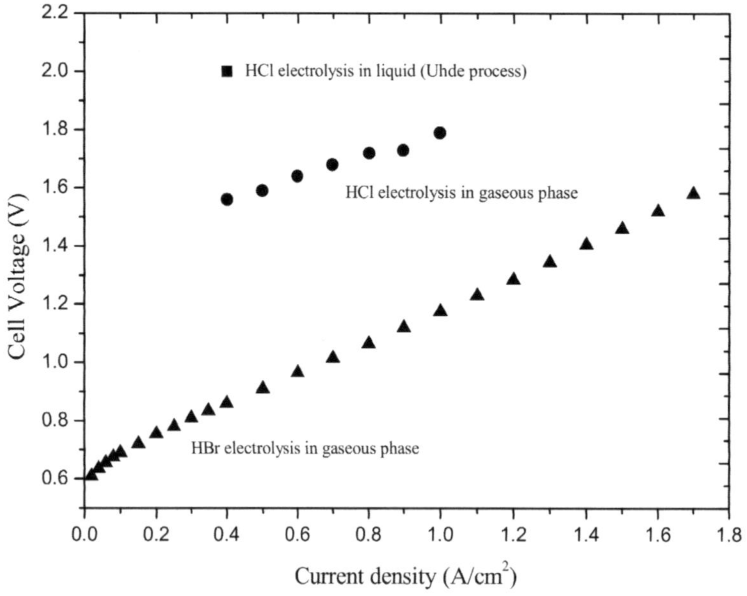

Figure 1: Comparison of the HBr process with HCl process with standard MEA.

Effects of the anode catalysts are in figure 2. It is seen that the non Pt catalyst of RuO_2, Vulcane XC 72R and TiO_2-Nb have comparable performance with the commercial MEA at the anode side.

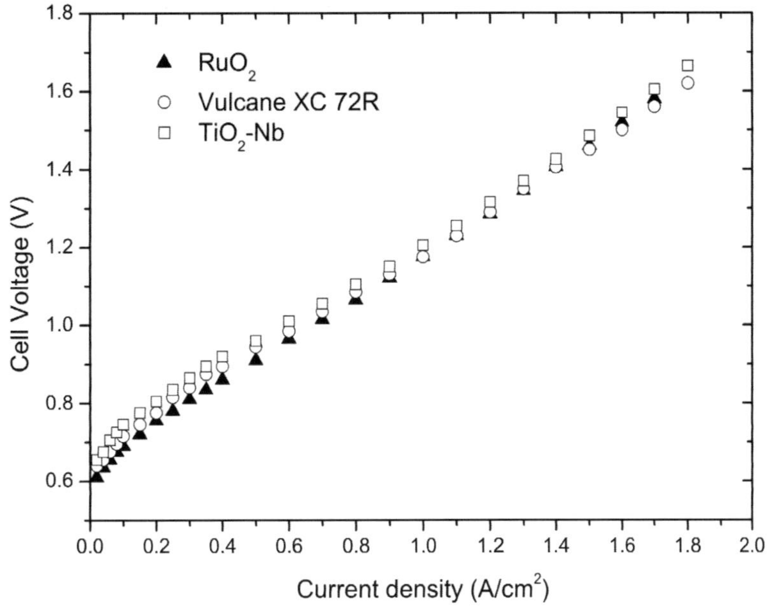

Figure 2: Effects of MEAs with different anode catalysts

From the previous results, it is seen that the overpotential for the HER catalyst takes a small part in the total cell voltage loss. Also, Pt is well known to be poisoned by the strong adsorption of Br^- and Br_2. Therefore, it is desired to develop non Pt electrocatalyst on the MEA. In this study, MoS_2, RuO_2, and IrO_2 are investigated, with respect to the synthesis procedures, screening the activity on the RDE and characterization. The results show that the low temperature synthesis of MoS_2 can significantly increase its activity. Also, the catalyst of IrO_2/C prepared from the EG method has a very high activity, which is even close to Pt. Characterization shows that the carbon support and the oxidation state of Ir probably play the key roles in this catalyst system.

4. Conclusion

The gas phase electrolysis has been successfully demonstrated to be used to produce hydrogen. HBr electrolysis is expected to have facile kinetics and thus the catalyst is not crucial, which enables us to use less expensive material at the anode side. Cathodic overpotential is relatively small than other voltage losses and thus we can use non Pt material to lower the cost and avoid Br^- and Br_2 poisoning.

Reference:

1. P. Sivasubramanian, R. P. Ramasamy, F. J. Freire, C. E. Holland and J. W. Weidner, *Int. J. Hydrogen Energy*, 32, 463 (2007).
2. W. Kondo, S. Mizuta, Y. Oosawa, T. Kumagai and K. Fujii, *Bull. Chem. Soc. Jpn.*, 56, 2504 (1983).

3. W. Kondo, Y. Shimizu, N. Miura and N. Yamazoe, *Int. J. Hydrogen Energy,* 13, 345 (1988).

4. W. Kondo, Y. Shimizu, N. Miura, N. Yamazoe, CN. Wauters and J. Winnick, *AIChE* J., 44, 2144 (1998).

5. S. Ferro and A. DE. Battisiti, *Journal of Applied Electrochemistry,* 34, 981 (2004).

6. S. Ferro, C. Orsan and A. DE. Battisti, *Journal of Applied Electrochemistry,* 35, 273 (2005).

7. J. M. Orts, R. Gomez and J. M. Feliu, *Journal of Applied Electrochemistry,* 467, 11 (1999).

8. J. M. Orts, R. Gomez, J. M. Feliu and A. Aldaz, *J. Phys. Chem.,* 100, 2334 (1996).

9. J. M. Orts, R. Gomez, J. M. Feliu and A. Aldaz. *Langmuir*, 13, 3016 (1997).

10. P. Sivasubranmanian, R. P. Ramasamy, F. J. Freire, C. E. Holland and J. W. Weidner, *Int. J. Hydrogen Energy*, **32**, 463 (2007).

11. D. Duphil et al. *Nanotechnology.,* 15, 828, (2004).

(506c) Hydrogen Production Via Two - Step High Temperature Thermochemical Process Using Porous $MnFe_2O_4$ Material

Michael K. Opoku, Rajesh Shende and Jan A. Puszynski, Chemical and Biological Engineering, South Dakota School of Mines and Technology, Rapid City, SD

Porous $MnFe_2O_4$ material was synthesized and investigated as a potential candidate for thermochemical water splitting. The basic idea is that manganese-ferrite can split water molecules by extracting oxygen atoms and reversibly incorporating them into its lattices. $MnFe_2O_4$ was produced from manganese acetate, iron citrate, and sucrose precursors via a novel microwave assisted synthesis method which resulted in a pure crystalline material. DSC/TGA redox experiments indicated that synthesized $MnFe_2O_4$ can gain up to 3.34wt% in oxygen atmosphere at 850 °C and 1 atm. The results have also showed that these materials can be fully regenerated in nitrogen atmosphere at 950 °C and 1 atm. Based on the TGA analyses of oxidation and reduction in air and nitrogen respectively, it was estimated that 47.7 STPml/g-solid of H_2 might be generated in each cycle. Hydrogen production in multiple cycles via a two-step thermochemical process using packed-bed reactor filled with porous $MnFe_2O_4$ material will be presented. In addition, details of synthesis and characterization of this promising material will be presented.

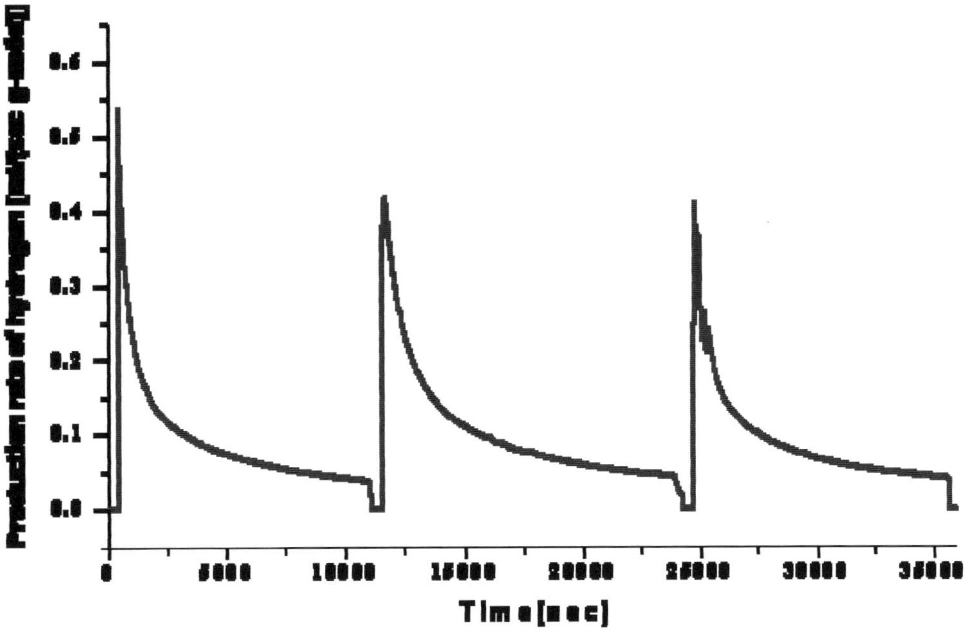

Figure 1. Production rate of hydrogen at 800 °C and atmospheric pressure using porous $MnFe_2O_4$ material.

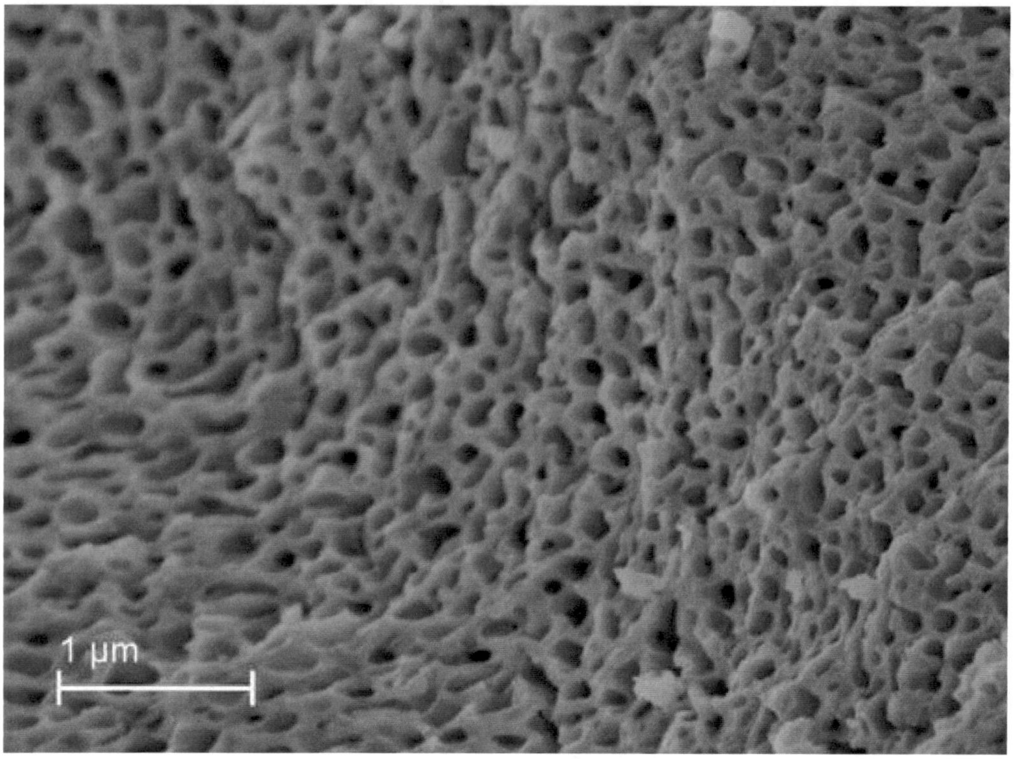

Figure 2. SEM image of synthesized $MnFe_2O_4$ materials.

See more of this Session: Advances in Thermochemical Hydrogen Production
See more of this Group/Topical: Topical 8: Hydrogen Production and Storage

H_2 GENERATION FROM THERMOCHEMICAL WATER-SPLITTING USING SOL-GEL SYNTHESIZED FERRITES ($M_xFe_yO_z$, M = Ni, Zn, Sn, Co, Mn, Ce)

Rahul R. Bhosale, Rajesh V. Shende and Jan A. Puszynski,

Department of Chemical and Biological Engineering,
South Dakota School of Mines & Technology,
Rapid City, SD 57701 USA

Abstract

This paper reports the sol-gel synthesis of Zn-ferrite with nanosized or porous morphology. This method involves the addition of chloride salts of Zn and Fe in ethanol followed by gelation using propylene oxide (PO). As-prepared gels were aged, dried and heated rapidly upto different temperatures of 600°C, 800°C and 1000°C and quenched in air or N_2 environment. Powder X-ray diffraction (XRD), scanning electron microscopy (SEM) and BET surface area analysis were performed to characterize the calcined ferrite powders. To achieve nanosize particles and higher specific surface area (SSA), the gels were rapidly calcined upto 600°C and quenched in air. The calcined Zn-ferrite powder thus obtained was then placed in an Inconel packed bed reactor and the H_2 generation and regeneration abilities were investigated by performing four consecutive thermochemical cycles where water-splitting and regeneration were carried out at 800°C and 1050°C, respectively.

Introduction

The energy density of H_2 (143 MJ/kg) is relatively higher than that of the fossil fuels such as oil (46.4 MJ/kg), gas (53.6 MJ/kg) and coal (32.5 MJ/kg), which makes it as an attractive and potential energy carrier. H_2 can be stored, delivered and converted to electricity by fuel cell technologies and used directly as a fuel without emitting any polluting gases such as SO_2 or carbon CO_2 [1]. Industrially H_2 is produced by gasification and reforming of fossil fuels, pyrolysis and reforming of biomass, and solar thermal processes [2]. However, by far the most attractive reaction for the production of H_2 is the thermal decomposition of water yielding pure H_2 [3]. Thermolysis of water (water-splitting in absence of chemical compounds) is extremely challenging as the direct thermal dissociation of water requires very high temperatures (>2500°C) and a unique high temperature membrane technology for separation of the H_2 and O_2 [4]. In contrast to this, thermochemical water-splitting cycles have huge potential as an environmentally and thermodynamically feasible approach for producing H_2 [5]. A metal oxide based thermochemical water-splitting is a two step process in which the first step belongs to the reduction of the metal oxide by releasing O_2 at higher temperatures; whereas, the second step corresponds to the oxidation of the reduced metal oxide by taking O_2 from water and producing H_2 via water-splitting at lower temperatures [2]. The combination of one regeneration / activation step (endothermic) and one water-splitting step (exothermic) is termed as one thermochemical cycle. The two-step thermochemical process eliminates the formation of explosive H_2-O_2 mixture, avoiding the need for high temperature separation of O_2 and H_2 [6].

Among the several redox materials investigated so far for the water-splitting application, ferrites such as $(A_xB_y)Fe_2O_4$ (doped spinel) or $A_xB_yFe_{1-x-y}O$ (doped wustite) or their mixtures with elements A and B being bivalent metal cations such as Mn, Ni, Zn, Co or Cu were reported to be most active materials for H_2 generation [4]. These ferrites have been synthesized by several synthesis methods such as solid-state mixing, oxidation of aqueous metal hydroxide suspension, self-propagation high temperature synthesis (SHS) and aerosol spray pyrolysis method [7 - 10]. Although beneficial, there have been reports indicating that these methods sometimes results into chemical heterogeneity, non-stoichiometry, coarser particles and lower specific surface area (SSA) [12]. It appears that coarser and non-porous particles may pose the hydrodynamic limitations such as diffusional and mass-transfer resistances if used in packed bed reactors. In addition to this, during the thermochemical cycles the redox materials undergo thermal cycling which might results into grain growth, thermal stress and structural disintegration such as spalling of the redox material coating. This may further clog the reactor and the process lines, which may results into increase in the pressure drop and thereby creating unsafe processing conditions [12].

We believe that the use of ferrite with porous nanoparticle morphology with higher SSA will lower down the water-splitting and regeneration temperatures. Additionally, it will help overcoming the diffusional limitations and mass-transfer resistances. The packed bed reactors loaded with porous ferrites with nanoparticle morphology will provide lower pressure drop and minimize resistance for gas transport. Ferrite with porous nanoparticle morphology can be synthesized using sol-gel synthesis method which involves molecular or atomic level mixing and networking of chemical components. By using the sol-gel method significant improvements in chemical homogeneity, stoichiometry, phase purity and powder characteristics, for instance, ultrafine powder and narrow particle size distribution are often realized [2]. The powders produced by the sol-gel synthesis method usually have higher SSA, which may provide higher number of active sites for water-splitting reactions. In our previous investigations we have reported the H_2 generation and regeneration abilities of sol-gel derived Ni-ferrite and Sn-doped Ni-ferrite at lower water-splitting and regeneration temperatures of 700°C and 900°C, respectively [2, 12]. The temperature 900°C is the lowest ever reported for the regeneration of the Ni-ferrite in thermochemical water-splitting application.

In this investigation, we report the synthesis of Zn-ferrite using sol-gel approch. As synthesized gels were aged, dried and calcined in air or N_2 environment. The effect of PO concentration and calcination temperature and environment on the compositional purity, crystallite size (C_s), SSA, pore volume (P_v) and pore diameter (P_d) were investigated in detail. The H_2 generation and regeneration ability of the calcined Zn-ferrite were investigated by performing four thermochemical cycles where water-splitting and regeneration were performed at 800°C and 1050°C, respectively.

Experimental Methods and Characterization

Zn-ferrite Synthesis

In this synthesis, anhydrous zinc chloride ($ZnCl_2$) and iron chloride tetrahydrate ($FeCl_2.4H_2O$) [1:0.95 (w/w) ratio] were dissolved in ethanol (C_2H_5OH) to obtain a clear solution. To this clear solution, a predetermined amount of propylene oxide (C_3H_6O) i.e. PO was added and the solution was set aside undisturbed at room temperature until a gel was observed. As-synthesized gels were aged for 24 h at room temperature, dried at 100°C for 1 h, and then heated rapidly (ramp of 40°C/min) upto various temperatures and quenched in air or N_2 environment.

Zn-ferrite Characterization

The compositional purity and C_s of the calcined Zn-ferrite powders at different temperatures were determined using Rigaku Ultima-Plus X-ray diffractometer (XRD) provided with Cu$K\alpha$ radiation (λ = 1.5406 Å, 40 kV, 40 mA) and a graphite monochromator. The Zeiss Supra 40VP field-emission scanning electron microscopy (SEM) was used to analyze the morphology of the calcined powders. BET surface area analyzer, Gemini II-2375 from Micromeritics, was used to determine the SSA, P_d and P_v of the calcined powders from the adsorption isotherm obtained after degassing the material at 200°C.

H_2 Generation Set-up

The water-splitting and regeneration ability of the sol-gel synthesized Zn-ferrite was investigated using the experimental set-up shown in Fig. 1. It consists of a tubular Inconel packed bed reactor enclosed in a vertical split furnace (Carbolite Inc., USA). The temperature of the furnace was controlled and regulated precisely with the in-built PID controller. Accurately weighed 5.0 g of the Zn-ferrite was packed in the middle section of the tubular reactor using raschig rings and quartz wool as support materials. A horizontal tube furnace was provided with a stainless steel tube and it was used as a water vaporizing/preheating set-up. Distilled water from the reservoir was fed to the vaporizing/preheating furnace using a metering pump (Fluid Metering Inc., USA) at the flow rate of 1 ml/min. Water vaporized and the preheated steam obtained was fed to the packed bed reactor with N_2 as a carrier gas. A mass flow meter for N_2 (AALBORG Inc., USA) was mounted on the feed line to control the carrier gas flow rate; a flow rate of about 100 SCCM/min was randomly selected. The exit gas stream from the packed bed reactor was first cooled down in a water bath and finally fed to a moisture adsorption column containing Drierite (anhydrous calcium sulfate), purchased from W.A. Hammond Drierite Company Inc., OH, USA. The exit gas stream (mixture of N_2 and H_2) from the adsorption column was continuously monitored using an online H_2 sensor, HY-OPTIMA 730 from H_2 Scan Corp., USA. The sensor provided H_2 concentration in vol % at an interval of 1 sec. For each water-splitting cycle, the water supply was shut down when the online H_2 sensor indicated a minimum of 0.5 (vol %) concentration level. Finally, the effluent gas stream was burned continuously.

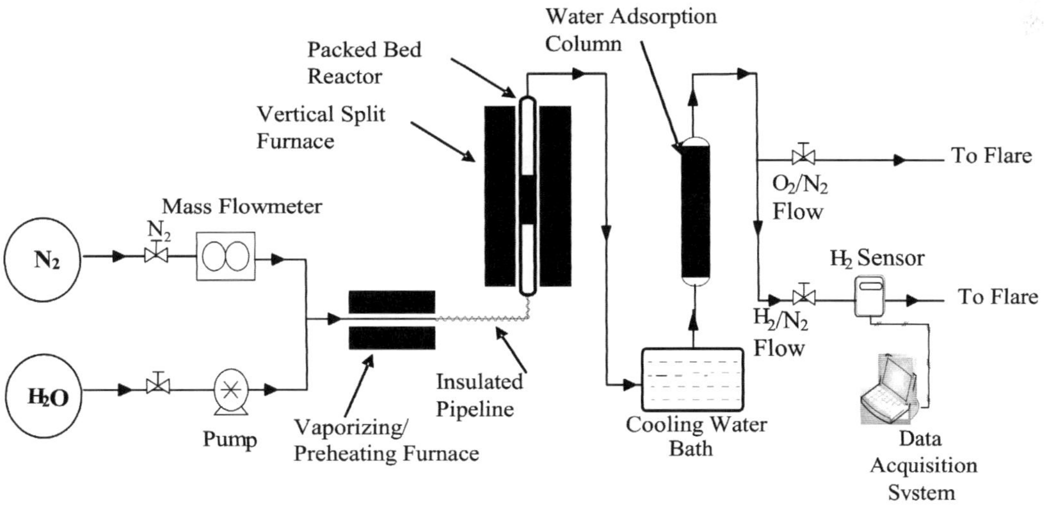

Fig. 1 Thermochemical water-splitting reactor set-up for H_2 generation [2].

Results and Discussion

Effect of Sol-gel Reaction Parameters and Calcination Temperature and Environment

In sol-gel synthesis method, addition of a proton scavenger such as PO helps in freezing the molecular chains and reduces the time required for the gel formation [2]. A set of experiments were performed to study the effect of PO concentration on gel time. To synthesize Zinc-ferrite, 10 g of the respective precursors (the weight ratios are mentioned in previous section) were added in 20 mL of ethanol and sonicated to obtain a clear solution. Different amounts of PO were added to achieve the gel formation. Table 1 presents the gel time with respect to PO addition. It was observed that as the concentration of PO increases, the time required for gel formation decreases. As-synthesized gels were aged for 24 h, dried at 100°C and heated rapidly upto 700°C and quenched in air or N_2 environment. The C_s, SSA, P_v and P_d of the calcined powders were determined and are presented in Table 1. These results indicate that the SSA and P_v of the calcined powders increases as the concentration of PO increases, whereas the C_s and P_d remain constant. The phase compositions of the calcined Zn-ferrite powders were observed to be unchanged irrespective of the PO concentration used.

Table 1. Effect of PO concentration on gel time, SSA, P_v, P_d and C_s of the sol-gel synthesized Zn-ferrite.

PO (mL)	Gel Time (sec)	Air				N_2			
		SSA (m^2/g)	P_v (cm^3/g)	P_d (nm)	C_s (nm)	SSA (m^2/g)	P_v (cm^3/g)	P_d (nm)	C_s (nm)
5	399	6.66	0.0089	5.393	30.33	5.49	0.00516	2.659	42.32
10	312	13.12	0.0201	5.424	29.83	10.45	0.00698	2.646	41.66
15	240	19.96	0.0278	5.432	29.44	16.33	0.01152	2.655	42.12
20	150	20.11	0.0296	5.387	29.41	16.98	0.01193	2.639	43.19

To study the effect of calcination temperature and environment, as-prepared gels with 20 mL of PO were aged for 24 h, dried at 100°C and heated rapidly upto 600°C, 800°C and 1000°C and quenched in air or N_2 environment. The compositional puritiy of the calcined powders were determined using X-ray diffractometer and the XRD patterns obtained are shown in Fig. 2a and 2b. These patterns indicate that the gels calcined in air at different temperatures resulted in a mixed composition containing $ZnFe_2O_4$, Fe_2O_3 and ZnO (Fig. 2a). Whereas, the gels calcined in N_2 environment resulted in mixed composition of $ZnFe_2O_4$, FeO and ZnO (Fig. 2b).

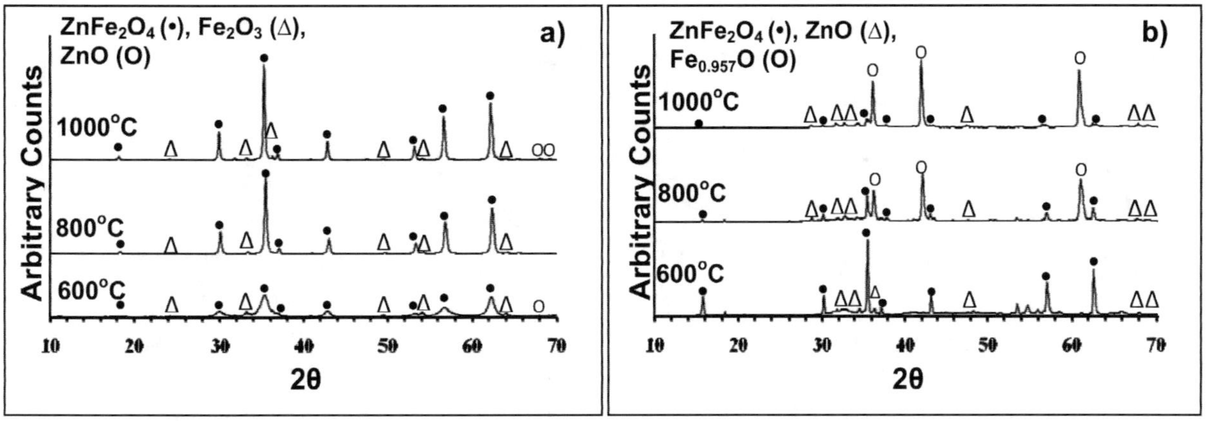

Fig. 2 XRD patterns of Zn-ferrites calcined at different temperatures in a) air and b) N_2 environments.

The BET SSA analysis results presented in Table 2. Indicate that the SSA, P_v, and P_d decrease and C_s increases with increase in calcination temperature. It can be observed that the samples calcined in air have higher SSA, P_v, and P_d as compared to that of calcined in N_2 environment. It was believed that the calcinations in N_2 environment might have induced the grain growth and decreased the SSA, P_v, and P_d, which required further validation using SEM analysis. The SEM images of calcined powders obtained after rapid calcination of the dried gels upto 600°C in air or N_2 environment are shown in Fig. 3a and 3b. The SEM image of the Zn-ferrite calcined in air indicated agglomeration of plate-like nanoparticles stacked together (Fig. 3a); whereas, the gels calcined upto 600°C and quenched in N_2 resulted into a foam- like

Table 2. BET and XRD analysis of the Zn-ferrites calcined at different temperatures in air or N_2 environment.

Calc. Temp. (°C)	Air					N_2				
	Phase Composition	SSA (m^2/g)	P_V (cm^3/g)	P_D (nm)	C_S (nm)	Phase Composition	SSA (m^2/g)	P_V (cm^3/g)	P_D (nm)	C_S (nm)
600	$ZnFe_2O_4$ (87%), Fe_2O_3 (12%), ZnO (1%)	36.43	0.0527	6.349	20.37	$ZnFe_2O_4$ (81%), FeO (6%), ZnO (13%)	21.33	0.0134	2.679	36.05
800	$ZnFe_2O_4$ (98%), Fe_2O_3 (2%)	13.66	0.0188	5.391	35.75	$ZnFe_2O_4$ (25%), $Fe_{0.957}O$ (68%), ZnO (7%)	11.78	0.0103	2.724	45.76
1000	$ZnFe_2O_4$ (95%), Fe_2O_3 (3%), ZnO (2%)	8.98	0.0115	5.098	40.53	$ZnFe_2O_4$ (6%), $Fe_{0.957}O$ (89%), ZnO (5%)	6.53	0.0058	2.623	50.92

morphology with few faceted larger particles of smaller (<1μm) and larger (>2μm) sizes (Fig. 3b). It indicates that the calcination in N_2 environment has produced few additional larger faceted grains, which were absent in the samples calcined in air.

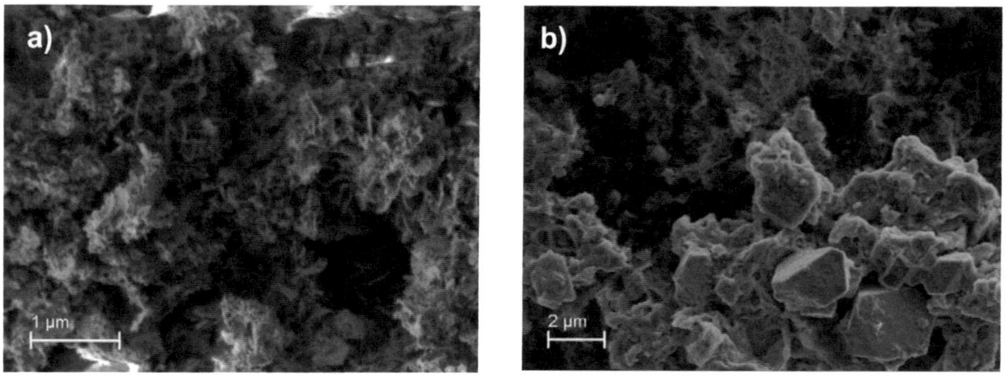

Fig. 3 SEM images of Zn-ferrite gels rapidly heated upto 600°C and quenched in a) air and b) N_2 environments.

H_2 Generation from Thermochemical Water-Splitting

Accurately weighed 5.0 g of Zn-ferrite was packed in the middle section of the Inconel tubular packed bed reactor and the temperature of the packed bed was increased upto the activation temperature of 1050°C. After activating the material at 1050°C for 4 h, the temperature of the packed bed was brought down to 800°C to perform water-splitting reaction. The preheated steam was fed to the packed bed reactor and the H_2 evolution was recorded continuously using the online H_2 analyzer. After water-splitting step, the material was regenerated at 1050°C for 4h in N_2 environment and the reactor again cooled down to 800°C to perform subsequent water-splitting cycle. Four consecutive thermochemical cycles were performed and the transient H_2 profiles obtained are shown in Fig. 4a. The results indicate that this material has regenerated itself at 1050°C, which is relatively lower regeneration temperature as compared with those that reported by other investigators. However, the H_2 levels were found to decrease in subsequent water-splitting cycle. From the H_2 profiles presented in Fig. 4a, the volume can be worked out using numerical integration method.

As the regeneration of the Zn-ferrite was performed at higher temperatures of 1050°C, Zn-ferrite might undergo thermal fatigue and that could eventually alter the powder characteristics. Therefore, after performing four thermochemical cycles, the ferrite material was removed from the reactor and analyzed using XRD, BET surface area analyzer and SEM. The XRD pattern of the reacted Zn-ferrite indicated a phase pure $ZnFe_2O_4$ (100wt %) indicating a good chemical stability during thermochemical cycling operation (Fig. 4b). The SSA of the original calcined Zn-ferrite before performing the water-splitting step was 36.43 m^2/g; whereas, for the material obtained after performing four thermochemical cycles it was decreased to 1.77 m^2/g. In addition to this, the P_v and P_d values of the reacted material were observed to be significantly lower as compared to that of original unreacted material. The SEM image shown in Fig. 4c support these results as the particle size was increased from nanosize to micron size with absence of any porous morphology. The investigation is in progress to examine the H_2 generation ability of Zn-ferrite material calcined in N_2 environment using thermochemical water-splitting.

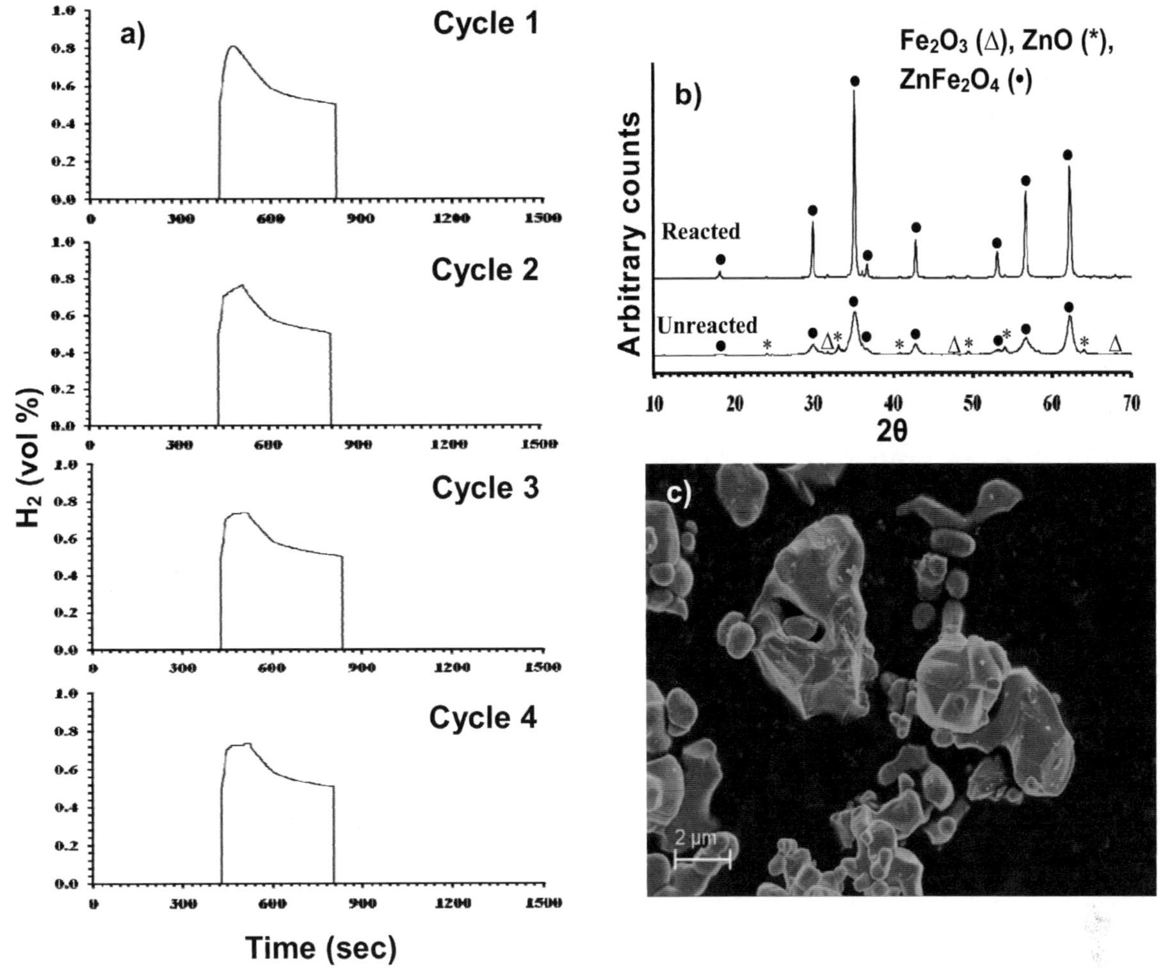

Fig. 4 a) Transient H_2 profiles obtained during four thermochemical cycles where water-splitting and regeneration were performed at 800°C and 1050°C, b) XRD patterns of the as-prepared Zn-ferrite and Zn-ferrite after four thermochemical cycles, c) SEM image of the Zn-ferrite after four thermochemical cycles .

Conclusions

Zn-ferrite material with nanosize or porous morphology was successfully synthesized using sol-gel method. The Effect of PO concentration and calcination temperature and environment on the compositional purity, SSA, C_s, P_v and P_d were studied in detail. Higher concentration of PO was found to reduce the gel time and increase the SSA of the Zn-ferrite. The results also indicated that the calcination temperature and environment have significant effects on the compositional purity and particle morphology. The Zn-ferrite gel rapidly calcined upto 600°C in air indicated a mixed phase composition containing the major phase as $ZnFe_2O_4$. The calcined Zn-ferrite powder was used for H_2 generation in four consecutive thermochemical cycles where water-splitting was performed at 800°C. The sol-gel synthesized Zn-ferrite material was observed to have potential to produce H_2 during water-splitting step at

800°C; however, its regeneration was not possible below 1050°C. The x-ray diffraction analysis of the reacted material revealed that the Zn-ferrite is chemically stable during thermochemical cycling operation. The BET SSA analysis results indicated that the SSA of the Zn-ferrite after performing four thermochemical cycles was significantly reduced. In addition to this, the SEM analysis of the reacted material showed that after four consecutive thermochemical cycles the ferrite results into non-porous micron size crystals.

Acknowledgement

The authors gratefully acknowledge the financial support by the National Science Foundation, grant award CBET-0756214.

References

1. Kang, K., Kim, C., Cho, W., Bae, K., Kim, S., and Park, C. (2009), "Novel two-step thermochemical cycle for hydrogen production from water using germanium oxide: KIER 4 thermochemical cycle," *International Journal of Hydrogen Energy*, **34**, pp. 4283-4290.
2. Bhosale, R. R., Shende, R. V. and Puszynski, J. A. (2010), "H_2 generation from thermochemical water-splitting using sol-gel derived Ni-ferrite" *Journal of Energy and Power Engineering*, **4**, pp. 27-38.
3. Roeb, M., Sattler, C., Kluser, R., Monnerie, N., Oliveira, L., Konstandopoulos, A., Agrafiotis, C., Zapalis, V., Nalbandian, L., Steele, A., and Stobbe, P. (2006), "Solar hydrogen production by a two-step cycle based on mixed iron oxides" *Journal of Solar Energy Engineering*, **128**, pp. 125-133.
4. Shende, R. V., Puszynski, J. A., Opoku, M. K., and Bhosale, R. R. (2009), "Synthesis of novel-ferrite foam material for water-splitting application," *Nanotech ISBN 978-1-4398-1782-7*, **1**, pp. 201-204.
5. Seralessandri, L., Varsano, F., La Barbera, A., and Padella, F. (2006), "On the oxygen-releasing step in the water-splitting thermochemical cycle by $MnFe_2O_4$ – Na_2CO_3 system," *Scripta Materialia*, **55**, pp. 875-877.
6. Bhosale, R. R., Shende, R. V., Opoku, M. K., and Puszynski, J. A. (2009), "H_2 generation from water-splitting using novel ferrite foam materials," *AIChE Conference Proceedings*.
7. Kojima, M., Sano, T., Wada, Y., Yamamoto, T., Tsuji, M., and Tamaura, Y. (1996). "Thermochemical decomposition of H_2O to H_2 on cation-excess ferrite," *J. Phys. Chem. Solids*, **57**, pp. 1757-1763.
8. Tamaura, Y., Hasegawa, N., Kojima, M., Ueda, Y., Amano, H., and Tsuji, M. (1998). "Water - splitting with the Mn(II)-ferrite-CaO-H_2O system at 1273K," *Energy*, **23**, pp. 879 – 886.
9. Agrafiotis, C., Pagkoura, C., Lorentzou, S., Kostoglou, M., and Konstandopoulos, A. (2007). "Hydrogen production in solar reactors," *Catalysis Today*, **127**, pp. 265 – 277.
10. Steinfeld, A. (2005). "Solar hydrogen production via a two-step water-splitting thermochemical cycle based on Zn / ZnO redox reactions," *International Journal of Hydrogen Energy*, **27**, pp. 611 – 619.
11. Bhosale, R. R., Shende, R. V., and Puszynski, J. A. (2010), "Sol-gel synthesis of ferrite foam materials for H_2 generation from water-splitting reaction," *Nanotech ISBN 978-1-4398-3401-5*, **1**, pp. 368 - 371.

10th AIChE Annual Meeting
November 7 – 12, 2010
Salt Lake City, Utah

Canada's Experimental Program on Nuclear Hydrogen Production with the Cu-Cl Cycle

G. F. Naterer
Faculty of Engineering and Applied Science
University of Ontario Institute of Technology (UOIT)
Oshawa, Ontario, Canada, L1H 7K4

ABSTRACT

This paper presents recent advances in Canada's program on nuclear hydrogen production with the thermochemical copper-chlorine (Cu-Cl) cycle. Enabling technologies for the Cu-Cl cycle are being developed by a Canadian consortium, as part of the Generation IV International Forum (GIF) for hydrogen production with the next generation of nuclear reactors. Recent advances towards an integrated lab-scale Cu-Cl cycle are discussed in this paper, including experimentation on the hydrolysis, spray drying, oxygen production and CuCl / HCl electrolysis processes.

1. INTRODUCTION

The ominous signs of climate change and air pollution are all around us: poorer air quality, retreating glaciers, illness and disease, stronger hurricanes, rising sea levels, among others. Hydrogen is a potentially major solution to the problem of climate change. But a key future challenge for hydrogen as a clean energy carrier is a sustainable, low-cost method of producing it in large capacities. Most of the world's hydrogen (about 97%) is currently derived from fossil fuels through some type of reforming process (such as steam-methane reforming; SMR). This paper presents recent advances in Canada's program of sustainable hydrogen production from nuclear energy, for a carbon-free source of hydrogen production.

Hydrogen is becoming an increasingly important energy carrier. It can be stored and used to generate electricity, either onboard vehicles or stationary power systems [1]. Hydrogen is also used widely by petrochemical, agricultural (e.g., ammonia for fertilizers), manufacturing, food processing, electronics, plastics, metallurgical, aerospace and other industries. For example, in Alberta, Canada, the oil sands sector needs vast amounts of hydrogen to upgrade bitumen to synthetic crude oil and remove impurities. Hydrogen has recently made significant commercial penetration with fuel cells in the lift-truck market, as well as back-up power systems for cell phone communication towers. The envisioned future hydrogen economy in the transportation sector will require significant increases to the world's current capacity to generate hydrogen, particularly in a clean, sustainable manner without relying on fossil fuels.

Thermochemical water decomposition is an emerging technology for sustainable large-scale production of hydrogen. Typically using two or more intermediate compounds, a sequence of chemical and physical processes split water into hydrogen and oxygen, without releasing any pollutants externally to the atmosphere. These intermediate compounds are recycled internally within a closed loop. Naterer et al. [2] presented an overview of Canada's program on nuclear hydrogen production with the thermochemical Cu-Cl cycle. Previous studies have identified over 200 possible thermochemical cycles [3, 4]. The sulfur-iodine (S-I) and hybrid sulfur cycles are prominent cycles that have been developed extensively by several countries. These include the USA (General Atomics, Savannah River National Laboratory, among others), Japan (Japan Atomic Energy Agency, JAEA), France (CEA), Italy and others [5, 6]. The following seven cycles (in addition to the S-I and hybrid sulfur cycles) were identified in a Nuclear Hydrogen Initiative [4] as the most promising cycles: copper-chlorine (Cu-Cl) [7], cerium-chlorine (Ce-Cl) [8], iron-chlorine (Fe-Cl) [8], vanadium-chlorine (V-Cl) [8], copper-sulfate (Cu-SO4) [8] and hybrid chlorine [8].

Experimental work has been conducted for processes in these cycles to demonstrate their scientific feasibility. However, most of these cycles require heat at temperatures over 800 C from very high temperature (Generation IV) reactors, which are not currently available and entail major design and material challenges. The Cu-Cl cycle has a significant advantage over these other cycles, due to lower temperature requirements around 530 C and lower. This paper focuses on recent experimental developments in Canada on the Cu-Cl cycle. Particular emphasis is given to hydrogen production with Canada's Super-Critical Water Reactor, SCWR.

2. THERMOCHEMICAL CU-CL CYCLE

The Cu-Cl cycle performs a sequence of reactions to achieve the overall splitting of water into hydrogen and oxygen. Using intermediate copper chloride compounds, the cycle decomposes water into hydrogen and oxygen, in a closed internal loop that re-cycles all chemicals on a continuous basis. Process steps in the Cu-Cl cycle and a schematic realization of the cycle are shown in Figure 1.

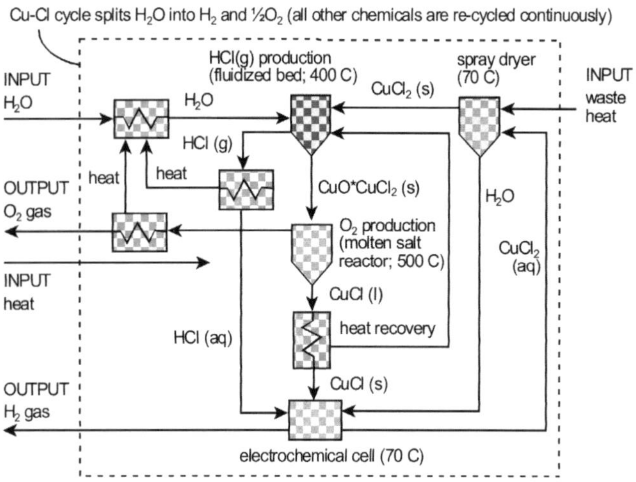

Figure 1: Schematic of the Cu-Cl cycle

There are three key variations of the Cu-Cl cycle: 5-step, 4-step and 3-step cycles. In the 5-step cycle, copper is produced electrolytically, moved to an exothermic thermochemical hydrogen reactor and then reacted with HCl gas to produce hydrogen gas and molten CuCl. The 4-step cycle combines these processes together to eliminate the intermediate production and handling of copper solids, through a CuCl / HCl electrolyzer that produces hydrogen electrolytically and aqueous Cu(II) chloride. Afterwards, the aqueous product is dried to generate Cu(II) chloride particles; then fed to a hydrolysis reactor to produce copper oxychloride. The 3-step cycle further combines these processes by supplying aqueous Cu(II) chloride directly into the hydrolysis chamber, such as spraying the solution with co-flowing steam to produce the same copper oxychloride product. This paper will focus on the 4-step cycle (Figure 1), since separation of the hydrolysis and drying processes can significantly improve the cycle's thermal efficiency.

3. CUCL / HCL ELECTROLYSIS FOR HYDROGEN PRODUCTION (STEP 1)

Oxidation of Cu(I) chloride (CuCl) during an electrochemical reaction occurs in the presence of hydrochloric acid (HCl) to generate hydrogen. The Cu(I) ion is oxidized to Cu(II) at the anode, and the hydrogen ion is reduced at the cathode. While the majority of the Cu species present is expected to be anionic (which would not permeate through the membrane), an appreciable concentration of cationic or neutral Cu species still exists and can permeate through the membrane and enter the cathode compartment where they are reduced to metallic copper. Metallic copper at the cathode acts as a poison, reducing the activity of the Pt catalyst towards the hydrogen evolution reaction (HER). This could result in a decrease in cell performance, so active research is underway to eliminate any possible copper crossover.

Atomic Energy of Canada Limited (AECL) has successfully demonstrated experimentally the production of hydrogen over a period of several days with a CuCl / HCl electrolyser. Hydrogen is produced electrolytically at the cathode and Cu(I) chloride is oxidized to Cu(II) chloride at the anode. The overall cell reaction can be expressed as

$$2CuCl + 2HCl \rightarrow 2CuCl_2 + H_2 \qquad (1)$$

At the anode, several Cu(I) and Cu(II) chloride complexation species can form, via the following reactions:

$$Cu^+(aq) + n\ Cl^-(aq) \leftrightarrow CuCl_n^{1-n} \qquad (2)$$

$$Cu^{2+}(aq) + n\ Cl^{-}(aq) \leftrightarrow CuCl_n^{2-n} \quad (3)$$

In chloride media, Cu(I) can exist as several different anionic species (e.g.$CuCl_2^-$, $CuCl_3^{2-}$). The relative ratios of each species are dependent on the concentration of chloride ions. Regardless of the exact species present, the anodic reaction proceeds by the electro-oxidation of Cu(I) species into Cu(II) species.

Recent studies have shown that the rate of the CuCl / HCl electrolysis reaction increases with temperature and CuCl concentration. A variety of electrode materials have been tested and hydrogen production was consistently observed at potentials as low as 0.5 V. However, approximately 0.65 V was typically required to achieve good current density. Promising results have successfully demonstrated that the reaction is feasible at low potentials and with relatively inexpensive electrode materials.

4. DRYING OF AQUEOUS CU(II) CHLORIDE (STEP 2)

Step 2 of the Cu-Cl cycle is the drying step as expressed by: $2CuCl_2\ (aq) \rightarrow 2CuCl_2\ (s)$. Aqueous $CuCl_2$ exiting from the previous electrolysis cell is dried to produce solid $CuCl_2(s)$, which is required for a subsequent hydrolysis step that produces copper oxychloride ($CuO*CuCl_2$) and HCl gas. The drying process is an energy-intensive step within the Cu-Cl cycle. Although the amount of heat required for the drying step is much higher than other steps in the cycle, it occurs at a lower temperature (lower quality) and therefore with heat that is more readily available. Within the drying step, the energy requirement increases from 1 to 5 times higher for slurry feed to solution, respectively, depending on the $CuCl_2$ concentration. The required heat can be obtained from low-grade "waste" heat from nuclear or other thermal power plants to improve the cycle efficiency. Spray drying is an efficient method of water removal due to the relatively large surface area available for heat and mass transfer, provided the liquid atomizes into sufficiently small droplets (on the order of a few hundred microns).

Experimental spray drying studies have been performed to produce solid Cu(II) chloride particles from an aqueous solution. Spray drying is a well established industrial process that allows powder characteristics and properties to be controlled and maintained nearly constant throughout a drying operation. Current research is performing experimental studies at successively lower inlet air temperatures, in order to eventually utilize waste heat from thermal power plants or other industrial sources for the spray drying heat requirements. Numerical studies have also been conducted to analyse the process of spray drying of aqueous Cu(II) chloride droplets. The results indicate that evaporative drying is possible down to low temperatures as low as 35°C, although such low temperatures may limit the product throughput rate.

5. HYDROLYSIS REACTION (STEP 3)

The following hydrolysis reaction occurs within the Cu-Cl cycle:

$$H_2O(g) + 2CuCl_2(s) \rightarrow Cu_2OCl_2(s) + 2HCl(g) \quad (4)$$

The reaction is an endothermic non-catalytic gas-solid reaction that operates between 350 and 400 C. The solid feed to the hydrolysis reaction is Cu(II) chloride, from the dried $CuCl_2$ product of step 2. Aqueous Cu(II) chloride is dried in step 2 to produce $CuCl_2$ particles, which are then moved to the hydrolysis unit and reacted with superheated steam to produce copper oxychloride solid and hydrochloric gas. This section describes recent developments in both experimental and numerical studies of transport phenomena associated with the non-catalytic gas-solid reaction of hydrolysis of Cu(II) chloride.

Lewis et al. [9] have demonstrated experimentally the scientific practicality of the hydrolysis reaction to produce HCl gas and solid copper oxychloride. Experimental results and simulations showed a large excess of water was needed for complete conversion (up to 98%) of $CuCl_2$. Byproducts other than Cu_2OCl_2 may appear, i.e., CuCl and CuOHCl, although ongoing research is investigating how and why these byproducts appear.

For an efficient and low capital cost process, it is important to reduce the water consumption. Aspen Plus simulations predict that 100% yield can be obtained with steam to $CuCl_2$ molar ratio of 17 at 370 C.

Unlike a spray reactor (nebulizer) used by the Argonne National Laboratory for the hydrolysis reaction, UOIT experiments have separated the drying and reaction processes, thereby using separate steps 2 (drying) and 3 (gas-solid reaction in a fluidized bed for hydrolysis). At UOIT, high temperature fluidized bed experiments with actual Cu-Cl compounds are being investigated, as well as low temperature experiments with simulated working fluids (air and spherical beads) to examine fluid dynamics and heat transfer phenomena in the fluidized bed.

The experimental unit was designed similar to the high temperature hydrolysis reactor, both geometrically and dynamically, by matching the Archimedes number, Froude number, density ratio, particle to column diameter and particle size distribution. The solid to gas density ratio is matched for accurate scaling of the hydrodynamics in a bubbling or turbulent bed. The experimental data indicates the importance of the pressure drop on the fluidization quality at high temperatures. Variations of particle properties at high temperatures reduce the fluidization quality.

Additional numerical and experimental studies have been performed to investigate the transport phenomena associated with the gas-solid reaction of particles in the fluidized bed during hydrolysis of Cu(II) chloride [2]. The conversion of solids depends on the rate of reaction and residence time of a particle.

A shrinking-core model has been developed to estimate the reaction rate of the gas-solid reaction. When the diffusion of gaseous reactant into a particle is much faster than the chemical reaction, the solid reactant is consumed nearly uniformly throughout the particle. In this situation, a uniform-reaction model can be used. On the other hand, when the diffusion of gaseous reactant is much slower and it restricts the reaction zone to a thin layer that advances from the outer surface into the particle, then the shrinking-core model is adopted. As the mole fraction of steam in the gas increases, the time of solid conversion decreases. Numerical simulations indicated that the fraction of reacted core volume to the total particle volume increases over time, regardless of the resistance mode for mixed flow of solids.

6. OXYGEN PRODUCTION BY COPPER OXYCHLORIDE DECOMPOSITION (STEP 4)

The oxygen production step receives solid feed of copper oxychloride ($CuO*CuCl_2$) and produces O_2 gas and molten Cu(I) chloride. The decomposition reaction is given by $CuO*CuCl_2(s) = 2CuCl$ (molten) + $0.5O_2$ (gas) at 530 C. Copper oxychloride particles decompose into molten salt and oxygen gas bubbles. The reactant particles absorb decomposition heat from the surrounding molten bath. Gas species leaving the oxygen reactor include oxygen gas and potentially impurities of products from side reactions, such as CuCl vapor, chlorine gas, HCl gas (trace amount) and H_2O vapor (trace amount). The substances exiting the reactor are molten CuCl, potentially unreacted solid $CuCl_2$ from the upstream hydrolysis reaction, due to the incomplete decomposition of $CuCl_2$, as well as reactant particles entrained by the flow of molten CuCl.

Recent Canadian developments have demonstrated experimentally the processes in a large engineering scale reactor, with flow capacities of 2 kg/day of oxygen production, by decomposing about 21 kg/day of solid $CuO*CuCl_2$ feed that enters the reactor. A number of experimental and operational issues with the unit reactor are currently being investigated, such as aggregation into solid blocks of $CuO*CuCl_2$ when conveying and feeding solid particles. The aggregation may choke or clog the feeder and cause sudden spouting of particles. Also, embedded particles of $CuCl_2$ from the upstream hydrolysis reactor in $CuO*CuCl_2$ may lead to undesirable products and side reactions, such as $CuCl_2$ decomposing to CuCl and Cl_2 gas. If particles enter the reactor at a temperature lower than 430 C, a challenge arises with the presence of bubbles in the molten salt. Some CuCl vapor might condense and molten CuCl can solidify around the $CuO*CuCl_2$ particles. If an aggregation develops with particles, molten salt and bubbles, the contact area between a reactant particle and heating medium (molten CuCl) will decrease and the aggregations float along the surface of the molten salt. This deters the decomposition of reactant particles and potentially leads to choking of the reactor (a major safety concern). These issues and others are currently under investigation for the copper oxychloride decomposition experiments.

Additional studies have also been performed to effectively recover heat from molten CuCl exiting the copper oxychloride decomposer. A configuration with convective heat transfer between molten CuCl droplets and air in a counter-current spray flow heat exchanger has been developed. Recovering heat from molten CuCl is challenging due to the phase transformations of molten CuCl, as it cools from liquid to two different solid states. Various configurations of heat recovery from molten CuCl have been examined.

7. CONCLUSIONS

This paper has presented an overview of Canada's experimental program on nuclear hydrogen production with the thermochemical copper-chlorine (Cu-Cl) cycle. The Cu-Cl cycle was identified by Atomic Energy of Canada Limited, AECL (CRL; Chalk River Laboratories), as the most promising cycle for thermochemical hydrogen production with the Generation IV nuclear reactor, SCWR (Super-Critical Water Reactor). Objectives of Canada's nuclear hydrogen program are to develop commercially viable processes for producing hydrogen based on the thermochemical Cu-Cl cycle and water electrolysis. The Cu-Cl cycle has attractive features that should help to meet these goals, particularly a 550 C maximum temperature that reduces demands on materials, compared to higher temperature cycles. Also, the Cu-Cl cycle couples well with various heat sources, such as the SCWR, solar power tower or Na-cooled fast reactor. The yields are nearly 100% in the hydrolysis and copper oxychloride decomposition processes, without catalysts and no recycle streams in these reactions. Also, conceptual process designs use commercially practiced processes in industry.

ACKNOWLEDGEMENTS

Support of this research and assistance from Atomic Energy of Canada Limited, Ontario Research Excellence Fund, Natural Sciences and Engineering Research Council of Canada (NSERC), University Network of Excellence in Nuclear Engineering (UNENE) and the Canada Research Chairs (CRC) program are gratefully acknowledged.

REFERENCES

[1] Stevens MB, Fowler MW, Elkamel A, Elhedhli S. Macro-level optimized deployment of an electrolyser-based hydrogen refuelling infrastructure with demand growth. Engineering Optimization 2008; 40 (10): 955 - 967

[2] Naterer, G. F., Suppiah, S., Lewis, M., Gabriel, K., Dincer, I., Rosen, M. A., Fowler, M., Rizvi, G., Easton, E. B., Ikeda, B. M., Kaye, M. H., Lu, L., Pioro, I., Spekkens, P., Tremaine, P., Mostaghimi, J., Avsec, A., Jiang, J., "Recent Canadian Advances in Nuclear-Based Hydrogen Production and the Thermochemical Cu-Cl Cycle", International Journal of Hydrogen Energy, vol. 34, pp. 2901 – 2917, 2009

[3] McQuillan BW, Brown LC, Besenbruch GE, Tolman R, Cramer T, Russ BE, Vermillion BA, Earl B, Hsieh HT, Chen Y, Kwan K, Diver R, Siegal N, Weimer A, Perkins C, Lewandowski A. High Efficiency Generation of Hydrogen Fuels Using Solar Thermochemical Splitting of Water: Annual Report, GA-A24972, General Atomics, San Diego, CA, 2002

[4] Lewis M, Taylor A. High Temperature Thermochemical Processes, DOE Hydrogen Program, Annual Progress Report, Washington, DC, 2006, p 182 - 185

[5] Sakurai M, Nakajima H, Amir R, Onuki K, Shimizu S. Experimental Study on Side-Reaction Occurrence Condition in the Iodine-Sulfur Thermochemical Hydrogen Production Process. International Journal of Hydrogen Energy 2000; 23: 613 – 619

[6] Schultz K, Thermochemical Production of Hydrogen from Solar and Nuclear Energy, Technical Report for the Stanford Global Climate and Energy Project, General Atomics, San Diego, CA, 2003

[7] Sadhankar RR, Li J, Li H, Ryland D, Suppiah S, Hydrogen Generation Using High-Temperature Nuclear Reactors, 55th Canadian Chemical Engineering Conference, Toronto, Ontario, October, 2005

[8] Carty RH, Mazumder M, Schreider JD, Panborn JB. Thermochemical Hydrogen Production, vol. 1, Gas Research Institute for the Institute of Gas Technology, GRI Report 80-0023, Chicago, IL 60616, 1981

[9] Lewis MA, Masin JG, Vilim RB, Serban M. Development of the Low Temperature Cu-Cl Thermochemical Cycle, International Congress on Advances in Nuclear Power Plants, May 15 - 19, 2005, Seoul, Korea

(618a) Reaction Engineering of a Novel Sulfur-Sulfur Thermochemical Water-Splitting Cycle

Alex Yokochi and Nicholas AuYeung, Department of Chemical Engineering, Oregon State University, Corvallis, OR

The Sulfur-Iodine and hybrid (Westinghouse) water-splitting cycles have suffered from the use of water as reaction media in their low temperature reactions. Separation of the various species from an aqueous medium results in significant energy expenditures and materials challenges in practical implementation. For the past few years we have examined an alternative route where the oxidation of sulfur step takes place in an ionic liquid medium. The application of ionic liquid reaction media for the low temperature reactions presents possible solutions to the problems of excess water removal, I_2 solidification, and SO_2 solubility.

The observation that carrying out the Bunsen reaction in ionic liquid media (Eq. 1 below) can be tailored to produce significant amounts of H_2S gas from the reaction mixture has led to the development of a novel Sulfur-Sulfur cycle, shown below.

$$4SO_2 + 4I_2 + 8H_2O \rightarrow 4H_2SO_4 + 8HI \quad \text{Eq. 1}$$

$$H_2SO_4 + 8HI \rightarrow H_2S + 4I_2 + 4H_2O \quad \text{Eq. 2}$$

$$\text{Net Low Temperature Reaction: } 4SO_2 + 4H_2O \rightarrow H_2S + 3H_2SO_4 \quad \text{Eq. 3}$$

$$H_2S + 2H_2O \rightarrow SO_2 + 3H_2 \quad \text{Eq. 4}$$

$$3H_2SO_4 \rightarrow 3SO_2 + 3H_2O + 3/2O_2 \quad \text{Eq. 5}$$

The production of H_2S (Eq. 2) following the Bunsen reaction, a normally undesired side reaction, can be intentionally enhanced under the appropriate conditions (concentration of I_2 and water, temperature and reaction time). The I_2 is regenerated in this reaction without having to transport HI to a decomposition section, and is therefore required only in catalytic amounts. The H_2S spontaneously desorbs from the reaction medium and can then be steam reformed to produce H_2 and SO_2, (Eq. 4) the latter of which is recycled. The thermal decomposition of H_2SO_4 to produce O_2 and SO_2 (Eq. 5) is common to other

Sulfur based thermochemical cycles. The H_2S generation and steam reformation steps (Eq. 3 and 4) have been demonstrated in our lab.[1] An overall thermal efficiency assuming a steam reformation temperature of 1100 K is estimated to be 55%.

This presentation will focus on the parametric dependence of the H_2S production and steam reforming reactions on reactant concentrations, temperature, and residence time, and details of the estimated thermal efficiency calculations will be presented.

---

[1] A. Yokochi, N. AuYeung, "*A Proposed New Sulfur-Sulfur Thermochemical Cycle*" *2009 Meeting of the AIChE.*

See more of this Session: Reactions and Separations for Hydrogen Production
See more of this Group/Topical: Topical 8: Hydrogen Production and Storage

(618b) Simultaneous Solubilities of Oxygen and Sulphur Dioxide in Water: Thermodynamic Data for the Sulphur Family of Thermochemical Cycles

Rachael H. Elder[1], Moises Romero[2], Andrew C. Shaw[2] and Ray W. K. Allen[2], (1)Department of Chemical and Process Engineering, University of Sheffield, Sheffield, United Kingdom, (2)University of Sheffield, Sheffield, United Kingdom

The Sulphur family of thermochemical cycles shows great promise for the massive scale production of hydrogen from water. The decomposition of sulphuric acid is a common stage in the cycles and poses design problems due to the corrosive nature of the solution, the high temperatures involved and the difficult separations required. The work considered here focuses on the oxygen separation stage of the process where, in order for a science based design approach to be taken, thermodynamic data are required for multicomponent phase equilibrium relationships between water, sulphur dioxide and oxygen; H2O-SO2-O2. A bespoke vapour-liquid-equilibrium still is used to measure the simultaneous solubilities of the ternary solution. The results are presented, along with a comparison to a model based on weak electrolyte thermodynamics. The model can be used for flash calculations, necessary for the design of the separation equipment.

See more of this Session: Reactions and Separations for Hydrogen Production
See more of this Group/Topical: Topical 8: Hydrogen Production and Storage

(618c) Storage and Separation of Hydrogen by the Metal Steam Process

Markus Thaler, Viktor Hacker and Matthaeus Siebenhofer, Institute of Chemical Engineering and Environmental Technology, Graz University of Technology, Graz, Austria

Certain metals react with water to build a metal oxide. During the reaction the water molecule is split and hydrogen is released as by-product. Using this behavior metals in their elemental form can be applied as chemical hydrogen storage material. The hydrogen is produced on demand. Only metal and water have to be stored in separate vessels thus hydrogen is not present in its physical form during the storage time.

Also the separation of hydrogen in a gas mixture can be achieved via repeated oxidation (Eq. 1) and reduction (Eq. 2) of metals and metal oxides. Reductive species such as CO can be used for the reduction of the metal oxide to reach a higher yield of hydrogen (Eq.3).

Oxidation step: xMe + yH2O → MexOy + yH2 (Eq. 1)

Reduction step: MexOy + yH2 → xMe + yH2O (Eq. 2) MexOy + yCO → xMe + yCO2 (Eq. 3)

The product gas of the reaction of metals with steam is especially capable for polymer electrolyte fuel cells as the hydrogen is already humidified.

However, some of the metals react at room temperature, others need significantly higher temperatures. To evaluate the materials capable for the storage and separation of hydrogen thermodynamic and experimental validations of the different metals are conducted. Metals as hydrogen storage materials are compared with established energy storage systems and advantages and disadvantages are worked out. By means of thermogravimetric analysis selected metals are reduced with hydrogen and oxidized with steam to confirm the theoretical results. Kinetic parameters are derived from the measurements for system calculation.

See more of this Session: Reactions and Separations for Hydrogen Production
See more of this Group/Topical: Topical 8: Hydrogen Production and Storage

(618d) The Kinetics of H2-D2 Exchange Over PdCu Surfaces

Casey O'Brien[1], James Miller[1], Bryan Morreale[1], Andrew Gellman[1] and Petro Kondratyuk[2], (1) USDOE National Energy Technology Department, Pittsburgh, PA, (2)Dept. of Chemical Engineering, Carnegie Mellon University, Pittsburgh, PA

Very high purity H_2 can be produced from mixed gas streams, such as coal gasification syngas, by the use of dense Pd-based membranes which have the unique ability to dissociate molecular H_2 and allow only atomic H to permeate through its bulk. To improve Pd's mechanical properties and its resistance to minor gas-phase components that can cause deterioration of membrane performance, such as H_2S, Pd is often alloyed with other metals, including Cu.

In this work, we use the H_2-D_2 exchange reaction to probe the H_2 dissociation activity of clean and sulfur-contaminated Pd, Cu and PdCu alloy surfaces. Over fixed beds of clean, diced Pd foils, the rate of H_2-D_2 exchange is limited by the desorption rate of product HD; over clean Cu, the reaction is limited by the relatively large barrier to dissociative adsorption. PdCu alloys display intermediate behaviors.

To provide a basis for understanding membrane performance in environments typical of the separation application, we also present kinetic analyses of H_2-D_2 exchange over Pd and PdCu surfaces in the presence of H_2S. Our results contribute to the design basis for next generation purification devices.

See more of this Session: Reactions and Separations for Hydrogen Production
See more of this Group/Topical: Topical 8: Hydrogen Production and Storage

(618e) Thermo Catalytic Decomposition of Methane Into Hydrogen and Carbon Nanofibers On Ni and Fe/ SiO2 Catalyst

Sushil Saraswat and Kamal Pant, Chemical Engineering, Indian Institute of Technology, Delhi, New Delhi, India

Thermocatalytic decomposition of methane is one of the promising technology for the production of pure hydrogen which is desirable for PEM fuel cell application. Thermo Catalytic methane decomposition to produce COx- free hydrogen and carbon nanofibers was carried out over Ni/SiO2 and Fe/SiO2 catalysts in a fixed bed quartz reactor. Metal components were loaded on SiO2 support by the wet impregnation method. Effect of metals loading on the support was investigated by increasing the metals loading from 20- 70 %. The morphology of the fresh catalysts and the characteristics of the carbon produced were also studied on the basis of the results of BET, Chemisorption, XRD, TGA, SEM and TEM analysis. The catalytic decomposition reaction was carried out at 700 deg C and atmospheric pressure. More than 80% methane conversion was obtained on Ni/SiO2 & Fe/SiO2 catalysts at 700ªC. Scanning electron microscope examinations revealed that the deposited carbon formed was in the form of carbon nanofibers. XRD analysis shows that carbon nanofibers formed are of graphitic nature. Transmission electron microscopy (TEM) studies revealed that filamentous carbon was formed during the reaction.

See more of this Session: Reactions and Separations for Hydrogen Production
See more of this Group/Topical: Topical 8: Hydrogen Production and Storage

(638a) Degradation Studies of Solid Oxide Electrolysis Cells

Greg G. Tao[1], Bruce Butler[1], Niladri Dasgupta[1], Anil V. Virkar[1] and Manohar S. Sohal[2], (1)Materials & Systems Research, Inc., Salt Lake City, UT, (2)Idaho National Laboratory, Idaho Falls, ID

Hydrogen produced by solid oxide electrolysis cells (SOEC) from steam represents one of the currently available economical means of mass-producing hydrogen, which requires the least purification process comparing to steam methane reformation and coal gasification. SOEC modules with long life and minimal degradation will be critical to reducing the cost of hydrogen (CoH), including but not limited to capital costs and operating costs. A key factor in developing robust SOEC technology is to understand degradation mechanisms and to implement viable mitigation strategies.

Collaborating with Idaho National Laboratory, Materials and Systems Research Inc. has been investigating possible chemical and structural degradation mechanisms in SOECs. Cell materials with a range of compositions have been developed to improve long-term performance of hydrogen production. Negative electrode-supported SOEC cells and short stacks have been fabricated and evaluated under various operating conditions. In this talk, performance in both a cell and stack levels will be presented and discussed.

See more of this Session: Developments in Electrolytic Routes to Hydrogen I
See more of this Group/Topical: Topical 8: Hydrogen Production and Storage

(638b) Renewable Energy Storage by CO2 Recycling Using High Temperature Electrolysis

Joseph J. Hartvigsen[1], Lyman Frost[2], S. Elangovan[2], Carl M. Stoots[3], James E. O'Brien[3] and Stephen Herring[4], (1)Ceramatec, Inc., Salt Lake City, UT, (2)Ceramatec, Inc, Salt Lake City, UT, (3)Thermal Fluids and Heat Transfer Department, Idaho National Laboratory, Idaho Falls, ID, (4)Nuclear Science & Engineering, Idaho National Laboratory, Idaho Falls, ID

We have significantly greater renewable energy resources than can effectively be utilized by the electric grid due to intermittency and remoteness of the resources from large load centers. A further barrier to greater exploitation of renewable energy is that in some locales it may be competing with very low cost generation resources. For example, in many regions off-peak nuclear power is available at very low cost. Of much more urgent concern than electricity generation is the cost and assured availability of transportation fuels. Developing a means of storing intermittent renewable energy and off-peak nuclear energy as liquid transportation fuels offers improved economic prospects and a solution to the energy security of our transportation fuels while simultaneously reducing net CO_2 emissions.

Many countries are now considering taxing CO_2 emissions, which will increase the cost of products that are associated with those emissions. The most common method currently considered for dealing with CO_2 emissions is the capture of the gas, pressurization, and then sequestration in either rock formations or saline aquifers. This is relatively costly in both capital investment and operation of the equipment. Also, there is the possibility that this CO_2 will escape at some point in the future subjecting the CO_2 generators to a perpetual and unquantifiable liability.

Ceramatec and INL have been investigating an alternative approach that converts CO_2 into valuable transportation fuel. Using the solid oxide fuel cell materials set in conjunction with a non-fossil electric energy sources it is possible to generate synthesis gas (CO and H_2) and oxygen from CO_2 and H_2O. This high temperature co-electrolysis (HTCE) process can use both thermal and electric energy inputs to electrolyze the CO_2 and H_2O at 100% thermodynamic efficiency. This is possible because the high temperature co-electrolysis reactions are endothermic, and the heat generated by resistance in the electrochemical device is chemically recuperated in the process. We have demonstrated conversion of this CO_2 derived syngas to liquid fuels via the Fischer Tropsch process.

See more of this Session: Developments in Electrolytic Routes to Hydrogen I
See more of this Group/Topical: Topical 8: Hydrogen Production and Storage

(638c) Electrolysis and Regenerative Operation of the NASA Symmetrical SOFC

Thomas Cable[1], John A. Setlock[2] and Serene C. Farmer[2], (1)NASA Glenn Research Center, Cleveland, OH, (2)Ceramics Branch, NASA Glenn Research Center, Cleveland, OH

Both solid oxide fuel cells and Regenerative SOFCs (RSOFCs) are of interest to NASA for Aeronautics and Space applications. Recent studies performed for future, more electric commercial aircraft have evaluated SOFCs for APUs and have considered RSOFC systems. RSOFCs, based on H_2/O_2, would operate in fuel cell mode while taxiing to the gate and then would perform water electrolysis during flight. The pure O_2 generated in the electrolysis mode could be used for emergency needs. Space applications for RSOFCs are also envisioned, using H_2O for Lunar missions and CO_2 for Mars missions. Interest in H_2 production by DOE has lead to collaboration with DOE/INL with a focus on electrolysis of the hot steam effluent at nuclear power plants.

NASA Glenn Research Center (GRC) has been evaluating a new solid oxide fuel cell concept consisting of a porous YSZ support structure on both sides of a thin YSZ electrolyte, which is later filled with active electrode materials, making it a bi-electrode supported cell (BSC). BSC cells with infiltrated electrodes were evaluated in the regenerative mode of operation, in order to measure the electrochemical voltage efficiency (EVE); which is the ratio of power generated in fuel cell mode (P_{out}) to the required power for electrolysis (P_{in}), measured at constant current density. The highest voltage efficiency measured for a 5 cm OD cell was 90 % at a current density of 200 mA/cm^2 at 850°C. Details of the cell fabrication, testing and electrochemical performance will be presented.

See more of this Session: Developments in Electrolytic Routes to Hydrogen I
See more of this Group/Topical: Topical 8: Hydrogen Production and Storage

(638d) Development of a Hydrogen Home Fueling System

Greg G. Tao, Bruce Butler and Anil V. Virkar, Materials & Systems Research, Inc., Salt Lake City, UT

The use of clean-burning hydrogen for transportation can provide a critical transition in the world's economy, protects public health and environment, while reduces the impact of petroleum price fluctuations and the dependence on non-domestic energy sourcing. In addition, hydrogen production can be decentralized, sidestepping complications in distribution and storage.

Presently, hydrogen is produced via coal gasification, natural gas steam reforming, or water electrolysis for central or forecourt production. Meanwhile, due to the lack of large-scale hydrogen storage, distribution networks, delivery and dispensing infrastructures, remarkable costs from the production plant-gate to the end-consumers (fuel cell vehicle drivers) will be added to the overall cost of hydrogen. Alternatively, the development and adoption of hydrogen home fueling systems can reduce or eliminate the need for the hydrogen storage and distribution. The success of this alternative, however, will depend largely upon the development of economical hydrogen home production technologies, to meet the immediate and long term needs for the quick market adoption of hydrogen-based vehicles.

Sponsored by the U.S. Department of Energy under the Small Business Innovative Research / Small Business Technology Transfer program, Materials and Systems Research Inc. (MSRI) is investigating the development of a hydrogen home fueling system that incorporates the concept of combined hydrogen, heat, and power (CH2P) tri-generation directly using distributed natural gas. Unlike the conventional hydrogen production approaches, this CH2P system is built upon advanced solid oxide fuel-assisted electrolysis cell (SOFECs) and solid oxide fuel cell (SOFC) hybrid technology.

In this talk, MSRI's SOFEC-SOFC technology development will be presented, along with corresponding performance characteristics. A system analysis of a hydrogen home fueling system equipped with this advanced hybrid technology will also be presented, followed by discussions of process economics and merits of the technology adoption.

See more of this Session: Developments in Electrolytic Routes to Hydrogen I
See more of this Group/Topical: Topical 8: Hydrogen Production and Storage

(638e) High Temperature Solid-Oxide Electrolyzers: The Role of Materials in Long-Term Performance Stability

S. Elangovan[1], Joseph Hartvigsen[2], Dennis Larsen[2] and Insoo Bay[2], (1)Ceramatec, Inc, Salt Lake City, UT, (2)Electrochemical, Ceramatec, Inc, Salt Lake City, UT

The materials and fabrication technology that are used for high temperature solid oxide fuel cells (SOFC) are, in general, directly applicable to high temperature solid oxide electrolysis cells (SOEC). Ceramatec, in collaboration with Idaho National Laboratory under the DOE Nuclear Hydrogen Initiative has been evaluating SOEC cells and stacks using SOFC materials set developed at Ceramatec. While the cells and stacks showed good reversibility between the two modes, and the performance scaled as expected from short stacks to tall stacks, a significant difference in long-term stability was observed. The SOEC mode of operation suffered much higher performance degradation.

Long-term performance stability of SOEC is found to be largely materials related. Some potential causes for degradation are common to SOFC device operation while others are unique to SOEC mode of operation. The dominant degradation mechanism that appear unique to SOEC operation was related to interface delamination of the oxygen evolution electrode. Oxidation of metallic components in high temperature and high steam conditions in SOEC mode is particularly severe near the seal areas. Other causes that may be common to both modes are: deposition of species at electrochemical sites to cause electrode poisoning and phase instability of current distribution layer. Under a project funded by the Office of Naval Research, systematic evaluation of various materials was undertaken to address the stability issues. Some of the changes evaluated include the use of Sc-stabilized zirconia electrolyte, new electrode materials set, protective coatings on interconnects and alternative seal options. These materials changes are found to improve stack stability significantly.

See more of this Session: Developments in Electrolytic Routes to Hydrogen I
See more of this Group/Topical: Topical 8: Hydrogen Production and Storage

Studies on solid oxide electrolysis cells –aspect of synthetic hydrocarbon fuel production

Xiufu Sun*, Søren H. Jensen, Sune D. Ebbesen, Mogens Mogensen

Fuel Cells and Solid State Chemistry Division, Risø National Laboratory for Sustainable Energy, Technical University of Denmark, Frederiksborgvej 399, 4000 Roskilde, Denmark

*xisu@risoe.dtu.dk

Recently, there has been an increased focus on the development of carbon neutral renewable energy due to the energy crisis and environmental pollution concerns. Solar energy and wind power are considered to be most plentiful and promising renewable energies in future; however, these renewable energy sources are troubled by meteorological fluctuations and geological variations. Therefore, the development of high efficiency technology for energy storage becomes very necessary. In recent years, the use of solid oxide electrolysis cell (SOEC) technology for energy conversion and storage has attracted much attention. With SOECs, electrical energy can be converted to hydrogen via steam electrolysis. SOECs are also capable of co-electrolysis of CO_2 and H_2O to produce synthesis gas (H_2+CO), which can be subsequently catalyzed to various types of hydrocarbon fuels. The produced synthetic hydrocarbon fuels can be transported or consumed based on the current infrastructure [1-2]. If the electricity comes from renewable sources, this technology may play a very important role in the future energy system.

Production of synthetic hydrocarbon fuels such as methane, methanol, dimethyl ether (DME) or synthetic diesel from renewable energy via synthesis gas needs elevated pressure in order to achieve a high conversion of synthesis gas (or syngas in short form). For example, a conventional process for methanol synthesis with Cu/ZnO-Al_2O_3 catalyst is operated at 200-300 °C in 45-60 bar [3]. The typical Fischer-Tropsch hydrocarbon synthesis with metal catalyst is normally operated at 10-60 bar [4]. The storage and transportation of gaseous products also need pressurization.

Hartvigsen and coworkers reported liquid hydrocarbon by catalytic production from syngas that was produced from SOEC at 20 bar [5]. The electrochemical model for hydrogen production by cathode supported SOECs developed by Ni and co-workers [6] suggested that for current densities below 0.5A/cm^2 no pressurization is recommended, whereas when operated at current densities above

0.5A/cm^2, pressurization is preferred. Increased pressure can increase diffusion rate though higher gradients and thus decrease the concentration over-potential.

The recent SOEC test results at Risoe DTU demonstrated that stable performance of electrolysis stacks can be achieved at current densities up to -0.75 A/cm^2 during co-electrolysis [2]. This shows that a pressurized operation may be preferred. At Risoe DTU, a high pressure SOEC test system has been developed and built in order to prove the concept of making synthetic fuel with integrated high pressure co-electrolysis of H_2O and CO_2. Fig. 1 shows current density versus voltage (i-V) curves of SOEC tested at 750°C with 50 % H_2 +50 % H_2O [7]. ASR calculated based on i-V curves show that with the increase of pressure from 1 bar to 10 bar the ASR decreased about 20 %.

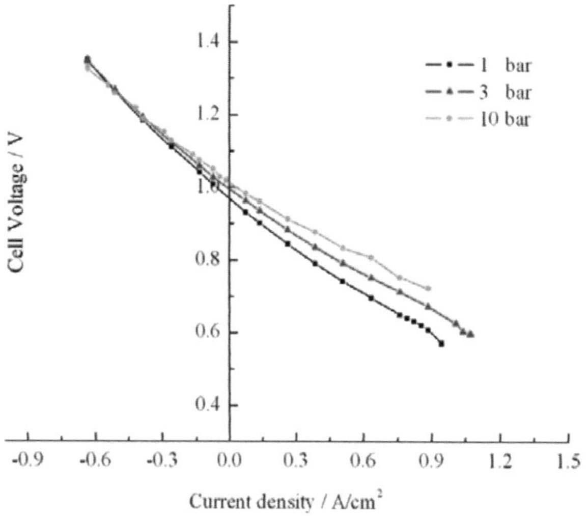

Fig.1 i-V curves for a Ni/YSZ based SOC at different pressure at 750°C with 50% H_2+50% H_2O supplied to the Ni electrode. Pure oxygen was supplied to the LSM/YSZ electrode. Positive current density means fuel cell mode and negative electrolyser mode.

For synthetic fuel production and SOECs performance improvement, pressurization has many advantages. Fig. 2 illustrates a system for synthetic hydrocarbon fuels production with high pressure SOECs. It is interesting to pressurize the SOECs, since less power is required to compress liquid water than to compress hydrogen. The pressurized H_2O and CO_2 enter a heat exchanger and the mixture is pre-heated before entering the SOEC stack. The typical SOEC operating temperature is around 700-900 °C. In the stack, H_2O and CO_2 are split to H_2 + CO at the cathode and O_2 at the anode. Subsequently, the syngas reacts catalytically into a synthetic hydrocarbon fuel. By choosing operation parameters such as pressure, H_2O/CO_2 ratio, catalyst, *etc*, the desired synthetic hydrocarbon fuel can be produced with such integrated system. Moreover, as shown in Fig. 3, it is possible to form CH_4 within the SOECs at high pressure, since the negative electrode of a SOEC is partly made of Ni, which is a good catalyst for CH_4 formation.

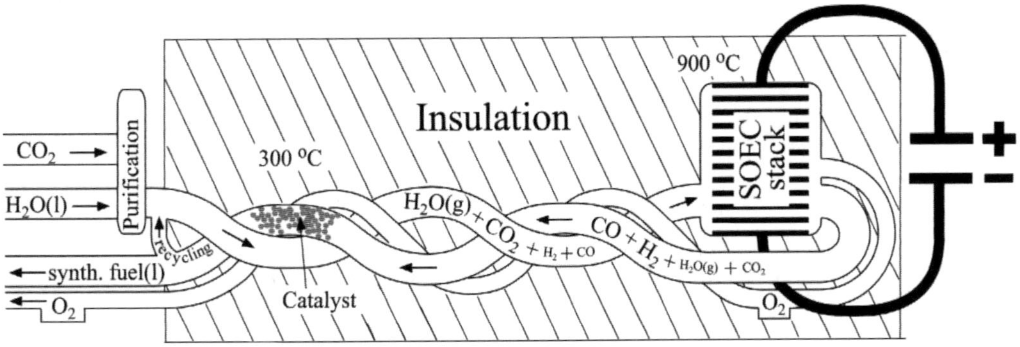

Fig.2 Sketch of a synthetic fuel production system integrated into a heat exchanger reactor coupled with high pressure co-electrolysis of H_2O and CO_2. The synthetic fuel is CO_2-neutral if CO_2 is captured from air and electricity used for the electrolysis reaction originates from renewable energy sources.

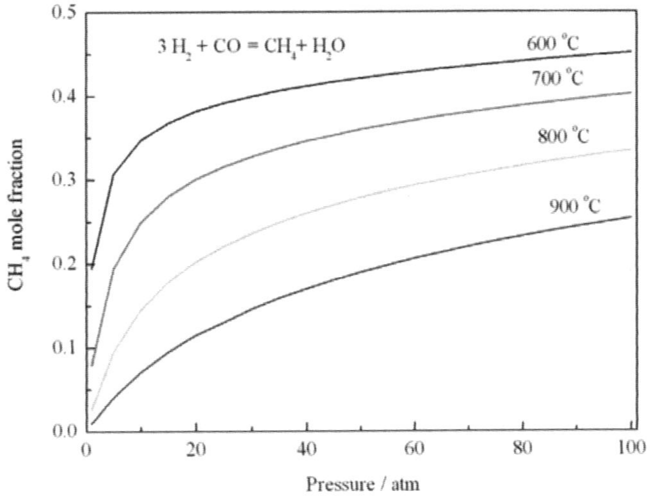

Fig.3 Equilibrium mole fraction of CH_4 formation from syngas as a function of pressure at 600,700,800 and 900°C, respectively.

However, as shown in Fig. 1 the open circuit voltage increases with increasing pressure. Therefore, the overall SOEC performance seems weekly affected by pressure. So far, theoretical studies of SOEC have focused on operation at 1 atm [6, 8-10], and to the best of our knowledge, no systematic theoretical modeling studies of SOEC operating in high pressure have been presented. A thermodynamic analysis of SOEC operating at increased pressure will be given in this presentation.

[1] Z. Zhan, W. Kobsiriphat, J.R. Wilson, M. Pillai, I. Kim, S.A. Barnett, Energy & Fuels, 2009, 23, 3089-3096.

[2] S. D. Ebbesen, M. Mogensen, Electrochem. Solid-State Letters, 2010, 13, B106-B108

[3] G.A. Mills, Fuel, 1994, 73(8) 1243-1279

[4] Gerard P. Van Der Laan, A.A.C.M. Beenackers, Catalysis Reviews, 1999, 41(3), 25-318.

[5] J. Hartivigsen, S. Elangovan, L. Frost, A. Nickens, C. Stoots, J. O'Brien, J.S. Herring, J. Electrochem.. Soc., 2008, 12(1), 625-637.

[6] M. Ni, M.K.H. Leung, D.Y.C. Leung, Int. J. Hydrogen Energy, 2008, 32, 2305-2313.

[7] S.H. Jensen, X. Sun, S.D. Ebbesen, R. Knibbe, M. Mogensen, Int. J. Hydrogen Energy, 2010, 35, 9544-9549.

[8] M. Ni, M.K.H. Leung, D.Y.C. Leung, J. Power Sources, 2006, 163, 460-466.

[9] J. Udagawa, P. Aguiar, N.P. Brandon, J. Power Sources, 2007, 166, 127-136.

[10] M. Liu, B. Yu, J. Xu, J. Chen, J. Power Sources, 2008, 177, 493-499.

(687b) Testing of the NASA Bi-Supported Cell Concept

Carl M. Stoots, Thermal Fluids and Heat Transfer Department, Idaho National Laboratory, Idaho Falls, ID, Robert O'Brien, Center for Space Nuclear Research, Idaho National Laboratory, Idaho Falls, ID and Thomas Cable, NASA Glenn Research Center, Cleveland, OH

The Idaho National Laboratory (INL) has been researching the application of solid-oxide fuel cell technology for large-scale hydrogen production. As a result, the INL has been testing various cell designs to characterize electrolytic performance. NASA, in conjunction with the University of Toledo, has developed a new cell concept with the goals of reduced weight and high power density for airborne power generation applications. The NASA cell architecture is planar and structurally symmetrical in that both electrodes support the thin electrolyte (called a bi-electrode supported cell, or BSC). The BSC design is of particular interest to the INL since delamination of the oxygen-side electrode has proven to be a problem in electrolyte-supported and hydrogen-electrode-supported cell designs operated as electrolyzers. This paper presents results of the INL's testing of this new solid oxide cell design as an electrolyzer. Time-series AC impedance spectroscopy was utilized to assess the long—term durability of this cell design and what factors primarily affect overall cell performance. Also to be discussed is the progress of NASA and INL to develop an all-ceramic stack from this concept.

See more of this Session: Developments in Electrolytic Routes to Hydrogen II
See more of this Group/Topical: Topical 8: Hydrogen Production and Storage

(687c) Experimental Evaluation of Electrode-Supported Solid-Oxide Cells for High-Temperature Electrolysis

James E. O'Brien, Thermal Fluids and Heat Transfer Department, Idaho National Laboratory, Idaho Falls, ID and Robert C. O'Brien, Thermal Science and Safety, Idaho National Laboratory, Idaho Falls, ID

Anode-supported cells represent some of the highest performing solid oxide fuel cells currently under development. The nickel cermet material, which serves as the anode in the fuel cell mode and the cathode in the electrolysis mode, has relatively high electronic conductivity and is therefore a logical choice for use as the mechanical support layer in electrode-supported cells. In an anode-supported SOFC, the anode is typically 1 – 1.5 mm in thickness while the electrolyte thickness can be as low as 10 μm. In the fuel-cell mode, steam diffusion away from the functional layer can be readily pressure-driven. In the electrolysis mode, however, it may be preferable to use an oxygen-electrode-supported cell to reduce the concentration overpotential associated with steam diffusion through the thickness of the electrode toward the functional layer. The typical oxygen-side electrode material for the present cells is lanthanum-strontium manganite (LSM) perovskite. Electrode-supported cells are currently being evaluated to determine their performance in the electrolysis mode. This paper will provide results of a experimental study of electrode-supported electrolysis cells. Comparisons will be made between performance of the cells in the fuel cell mode and in the electrolysis mode. The effect of steam fraction will be examined. The cells will also be characterized using impedance spectroscopy. Long-term degradation in the two modes of operation will also be evaluated.

See more of this Session: Developments in Electrolytic Routes to Hydrogen II
See more of this Group/Topical: Topical 8: Hydrogen Production and Storage

CFD MODEL OF ELECTRODE SUPPORTED PLANAR SOLID OXIDE ELECTROLYSIS CELLS

Grant Hawkes, James O'Brien
Idaho National Laboratory, Idaho Falls, Idaho

ABSTRACT

A three-dimensional computational fluid dynamics (CFD) electrochemical model has been created to assess high-temperature electrolysis performance of an electrode supported Solid Oxide Electrolysis Cell (ES-SOEC). Electrode-supported cells represent some of the highest performing solid oxide fuel cells currently under development. The nickel cermet material, which serves as the hydrogen electrode, (anode in the fuel cell mode and the cathode in the electrolysis mode) has a relatively high electronic conductivity and is therefore a logical choice for use as the mechanical support layer in electrode-supported cells. In an anode-supported SOFC, the anode is typically 1 – 1.5 mm in thickness while the electrolyte thickness can be as low as 10 μm. In the fuel-cell mode, steam diffusion away from the functional layer can be readily pressure-driven. In the electrolysis mode, however, it may be preferable to use an oxygen-electrode-supported cell to reduce the concentration overpotential associated with steam diffusion through the thickness of the electrode toward the functional layer. The typical oxygen-side electrode material for the present cells is lanthanum-strontium manganite (LSM) perovskite. Electrode-supported cells are currently being evaluated to determine their performance in the electrolysis mode. This paper will provide 3D CFD results of a computational study of electrode-supported electrolysis cells. The objective of the work is to determine the relative advantages and disadvantages of hydrogen electrode-supported cells versus oxygen electrode-supported cells for operation in the electrolysis mode. Computational results will be validated against experimental data obtained with hydrogen electrode-supported cells.

INTRODUCTION

A three-dimensional computational fluid dynamics (CFD) model has been created to model high-temperature steam electrolysis in a cross-flow planar solid oxide electrolysis cell (SOEC). The model represents a single cell, as it would exist in an electrolysis stack. Details of the model geometry are specific to two different cells, one fabricated by Julich and the other by St. Gobain[1].. The Julich cell is compared to literature [1] and the St. Gobain cell [2] is compared to experimental tests performed at the INL. Mass, momentum, energy, and species conservation and transport are provided via the core features of the commercial CFD code FLUENT[2]. A solid-oxide fuel cell (SOFC) module adds the electrochemical reactions and loss mechanisms and computation of the electric field throughout the cell. The FLUENT SOFC user-defined subroutine was modified for this work to allow for operation in the SOEC mode. Model results provide detailed profiles of temperature, Nernst potential, operating potential, activation over-potential, anode-side gas composition, cathode-side gas composition, current density and hydrogen production over a range of stack operating conditions. Mean model results are shown to compare favorably with experimental results obtained from single cell tested at INL.

Voltage versus current density curves for experimental and computational are presented for the Julich and St. Gobain cells. Discussion of thermal neutral voltage, enthalpy of reaction, hydrogen production, cell thermal efficiency, cell electrical efficiency, and Gibbs free energy are discussed and reported herein.

1,2 References herein to any specific commercial product, process, or service by trade name, trademark, manufacturer, or otherwise, does not necessarily constitute or imply its endorsement, recommendation, or favoring by the U.S. Government, any agency thereof, or any company affiliated with the Idaho National Laboratory

When predicting experimental results, empiricism arises. When trying to predict experimental results with this SOEC model, the only way to match the experimental data is to change some of the electrochemical parameters shown below.

NOMENCLATURE

ASR	area specific resistance, Ω-m^2
E	voltage potential, V
F	Faraday constant, 96487 J/V-gmol
ΔG	Gibbs free energy, J/gmol
ΔH	molar enthalpy of reaction, J/gmol
I	current, A
j	electrons transferred per H2 molecule
K	permeability, m^2
k	thermal conductivity, W/m^2-K
$\dot{N}$	molar flow rate, gmol/s
P	pressure in stack, Pa
Q	external heat transfer, W
R	universal gas constant, J/gmol-K
T	temperature, K
V	voltage, Volts
W	work, product of I*V, W
y	molar fraction

Greek Letters

φ	porosity
ρ	specific resistance Ω-m
η	efficiency

Subscripts

act	activation overpotential
$H2$	Hydrogen gas
$H2O$	steam
mf	mass fraction
o	open-cell
$O2$	Oxygen
R	reaction
std	standard pressure
tn	thermal neutral
t	thermal

NUMERICAL MODEL

The numerical model developed for this report is based on the geometry of a single SOEC cell designed and fabricated by St. Gobain and tested at the INL. The Julich cell has the same fabrication and dimensions. An illustration of this single cell experiment is shown in Figure 1. The stack has a per-cell active area of 16 cm^2. It is designed to operate in cross flow, with the steam/hydrogen gas mixture entering the inlet manifold on the bottom in the illustration.

The numerical model geometry represents a single cell, as it would exist the single cell experiment. The numerical domain extends from the bottom of the hydrogen side current collector to the top of the oxygen side current collector as shown in Figure 2. Included in the electrode portion of each electrode is a small layer called a functional layer that transitions from the electrode to the electrolyte. The H2 side functional layer is 8 microns, while the O2 side functional layer is 10 microns. These functional layers are tighter and have smaller porosity and smaller permeability with a larger tortuosity.

The FLUENT SOFC module treats the electrolyte as a 2-D planar element. Therefore the electrolyte in the model has geometrical thickness of zero. On either side of the electrolyte are the electrodes that are created with 3-D elements. Therefore, the electrolyte/electrode assembly in the model is only as thick as the two electrodes.

The numerical grid used in this study included 10 elements each in the flow inlet and outlet regions, and 30 x 30 in the active cell area in the X and Y directions and 26 elements through the thickness. Inlet and outlet flow regions were included in the model so that no numerical gradients would exist at the active cell area. Each flow channel (current collector) has 10 elements through its thickness, whereas each electrode has three elements through its thickness. Two views of the numerical grid used in the FLUENT model are shown in Figure 3. Figure 3(a) is a corner view showing 28 elements stacked in the z-direction, representing 5 distinct layers. Figure 3(b) is a top view showing the 42 x 42 element grid pattern used in the model. The center layers visible in Figure 3(a) represent the electrode regions. There are three elements for each electrode layer. Note that since the FLUENT SOFC module treats the electrolyte as a 2-D planar element, the electrolyte is not visible in Figure 3(a). Reference [4] shows that good agreement with as rough as a 5 x 5 grid, whereas this grid for this model is 30 x 30 for the electrolyte area.

Shown in Table 1 are the geometric and input parameters for the base case of this model. Parameters include thickness, thermal conductivity, specific resistivity, porosity, tortuosity and permeability.

Table 2 shows mass flow rates and species concentration for the Julich model and the St. Gobain model. Additional parameters specified in the numerical model include the electrode exchange current densities, mass flow rates, electrolyte ionic conductivity, gap electrical contact resistance, and concentration exponent used in activation overpotential.

All external surfaces are fixed at 1173 K. The gas flow inlets are specified in the FLUENT model as mass-flow inlets, with the gas inlet temperatures are set at 1173 K and the inlet gas composition determined by specification of the mass fraction of each component. The gas flow rates used in the model are shown in Table 2.

DISCUSSION

In the electrolysis mode, the net heat flux is negative at low current densities, increasing to zero at the "thermal-neutral" voltage, and positive at higher current densities. Assuming the process occurs at a specified temperature, the thermal-neutral voltage can be predicted from direct application of the First Law to the overall system:

$$Q - W = \dot{N}_{H2} \Delta H_R \qquad (1)$$

Letting $Q = 0$ (no external heat transfer), $W=VI$, and noting that the electrical current is directly related to the molar production rate of hydrogen by

$$\dot{N}_{H2} = I/2F \qquad (2)$$

where F is the Faraday number (F = 96,487 J/V-gmol), yields:

$$V_{tn} = -\Delta H_R / 2F \qquad (3)$$

Since the molar enthalpy of reaction, ΔH_R, is strictly a function of temperature (albeit a very weak function), the thermal-neutral voltage is also strictly a function of temperature, independent of cell ASR and gas compositions. The particular values of net cell heat flux at other operating voltages do however depend on cell ASR and gas compositions. The thermal-neutral voltage increases only slightly in magnitude over the typical operating temperature range for steam electrolysis cells, from 1.287 V at 800°C to 1.292 V at 1000°C. Stack operation at or below the thermal-neutral voltage simplifies thermal management of the stack since excess air flow is not required. In fact, in the electrolysis mode, since oxygen is being produced, there is also no theoretical need for air flow to support the reaction at all. In a large-scale electrolysis plant, the pure oxygen produced by the process could be saved as a valuable commodity. Careful consideration must be given, however, to the choice of materials for containing pure oxygen at elevated temperatures. In addition, it may be desirable to sweep with air or some other gas in order to minimize the effects of any hydrogen leakage.

A thermal efficiency, η_t, can be defined for electrolysis cells, analogous to the fuel cell efficiency definition presented in textbooks on fuel cells. The thermal efficiency quantifies the heating value of the hydrogen produced by electrolysis per unit of electrical energy consumed in the stack. Based on this definition,

$$\eta_t = \frac{\Delta H_R \dot{N}_{H2}}{VI} \qquad (4)$$

Eliminating the current I, the thermal efficiency can be expressed in terms of cell operating potential as:

$$\eta_t = \frac{\Delta H_R / 2F}{V} = \frac{V_{tn}}{V}. \qquad (5)$$

The thermal efficiency for the fuel-cell mode of operation is the inverse of Eqn. (5). It should be noted that the value of the thermal efficiency defined in this manner for electrolysis can exceed 1.0. As an example, for the reversible stoichiometric case, the cell potential approaches reference open-cell value, $E_o = \Delta G_R / 2F$, yielding:

$$\eta_{t,\max} = \frac{\Delta H_R}{\Delta G_R} \qquad (6)$$

which for steam electrolysis at 850°C is equal to 1.34. For cases with variable gas concentrations, the open-cell potential is given by the Nernst Equation, which for the hydrogen/oxygen/steam system takes the form:

$$E = E_o - \frac{RT}{jF} \ln\left[\left(\frac{y_{H2O}}{y_{H2} y_{O2}^{1/2}}\right)\left(\frac{P}{P_{std}}\right)^{-1/2}\right] \qquad (7)$$

FLUENT calculates the activation over-potential by the following set of equations. The parameter of 10000 and 1e8 A/m^2 used in the two different models is $i_{,0ref}$.

$$i_{0eff} = i_{0,ref} (Y_j)^{\gamma} \qquad (8)$$

Where (Y_j) is the mole fraction and γ is the concentration exponent set to 0.5. With i being the local current density, the activation potential for the cathode and anode can then be calculated as:

$$V_{act} = \frac{2RT}{jF} \sinh^{-1}\left(\frac{i}{i_{0eff}}\right) \qquad (9)$$

RESULTS

Results of the FLUENT simulations obtained for various cases are presented in Figures 4 through 14. Figures 4 through 7 show the results when comparing to the Julich cell in the fuel cell mode at 19.2 A. Figures 8 through 11 show the comparisons to the St. Gobain experiment conducted here at INL in the electrolysis mode at -10.0 A. Figures 12 through 14 show the numerical results when comparing cells that are hydrogen electrode-supported compared t oxygen electrode-supported. Figure 4 shows the current density contours for the fuel cell mode at 19.2 A. Steam/hydrogen flow is from left to right, while the air side flow is from top to bottom. Highest current densities occur at the top left of the model where the H2 and 02 concentrations are the highest and most favorable. Figure 5 shows contours for H2 mass fraction at 19.2 A. H2 is consumed in the fuel cell mode as it flows from left to right. Figure 6 shows a contour plot of the 02 mass fraction as it flows from top to bottom. The 02 is consumed as it flows through the cell in the fuel cell mode.

Figure 7 shows the V-i curve for the FLUENT comparisons to the Julich cell. A few data points are presented in the electrolysis area for the CFD model, but not in the experimental data. Several different variations are plotted also. An exchange current density reference value for the H2 electrode was input at 1e4 A/m^2. A contact resistance of 3.94e-6 Ohm-m^2 was also input into the model. A temperature boundary condition was input at 1077 K for the edges of the model simulating the oven temperature as a boundary condition. An offset exists for the entire set of data between the model and the experiment. The curvature is due to the species concentration in the Nernst potential. The curvature of the experiment and model are almost identical for the contact resistance model.

Figure 8 shows the current density when of the model for -10.0 A. The largest magnitude (negative value) occurs at the top left of the model where the H2 and 02 concentrations are the lowest and the H2O concentration is the highest. Figure 9 shows a contour plot of the mole fraction of H2 for the model at -10.0 A. H2 is being produced as it flows through the cell from left to right. Figure 10 shows the 02 concentration increase as 02 is produced from top to bottom.

Shown in Figure 11 is the V-i curve when comparing the St. Gobain experiment to the model. Both line plots of the model are for 85 kPa absolute pressure. The T70dp stands for the dew point at 70 °C for the bubbler to introduce the steam into the flow. Good agreement exists between the experiment and model.

Figures 12 through 14 show the V-i curve, temperature versus current, and functional layer pressure, respectively when comparing cells with a hydrogen electrode-supported cell and an oxygen electrode-supported cell. No discernible difference exists in the V-i curve or the temperature versus current plots. Shown in

Figure 14 are the pressure curves of the functional layers versus current for both types of cells. The blue and red lines show the pressure in the hydrogen electrode and oxygen electrode functional layers respectively for a hydrogen electrode supported cell. A minimum pressure of -9977 Pa at -10.0 A exists for the hydrogen electrode supported cell the functional layer. This pressure exists as the steam/hydrogen is being "sucked" into the reaction area in the electrolysis mode. The maximum pressure at -10.0 A for the oxygen electrode functional layer in this cell is 1817 Pa.

When reversing this cell with a thick oxygen electrode, the maximum pressure in the oxygen functional layer is 6137 Pa at -10.0 A, while the minimum pressure of the hydrogen functional layer is -2911 Pa at -10.0 A. All of these pressures are very small, but they do show the pressure differences in the cells for a reversed design.

CONCLUSIONS

A three-dimensional computational fluid dynamics (CFD) model has been created to model high-temperature steam electrolysis in a planar solid oxide electrolysis cell (SOEC). Comparisons between an experimental model from a Julich cell in the fuel cell mode taken from literature, and a St. Gobain cell in the electrolysis mode taken from an INL experiment have been presented. Good agreement exists between the experimental data and the model predictions for the operation voltage versus current density after varying electrochemical parameters to match the experimental data. Numerical predictions were also presented when reversing a cell with a thick oxygen electrode in the electrolysis and fuel cell mode. Cell performance did not change using this cell, however, pressure differences were noted to be lower for the reversed cell.

REFERENCES

1) Haanappel, V. A. C., et. al., "Characterisation of Ni-cermets SOFCs with varying anode densities.", Journal of Power Sources 171 (2007) 789-792.
2) O'Brien, J. E., Stoots, C. M., "High Temperature Electrolysis using Electrode Supported Cells.", paper # B506, European Fuel Cell Forum 2010, Lucerne Switzerland, July 2010.
3) FLUENT Users Manual, version 6.3.26, Lebanon New Hampshire, 2006.
4) Hawkes, G. L., O'Brien, J. E., Stoots, C. M., Herring, J. S., and Jones, R. W., "CFD Model of a Planar Solid Oxide Electrolysis Cell: Base Case and Variations," paper # HT2007-32310, 2007 ASME-JSME Thermal Engineering Conference and Summer Heat Transfer Conference, July 8 – 12, 2007, Vancouver, BC, Canada.

ACKNOWLEDGEMENTS

Work supported by the U.S. Department of Energy, NGNP Program, Idaho Operations Office Contract DE-AC07-05ID14517.

Table 1. Material properties and dimensions of models.

Material properties	
ρ O2 electrode/functional layer (Ω-m)	1.425 x 10^{-4}
ρ H2 electrode/functional layer (Ω-m)	8.856 x 10^{-6}
ρ electrolyte (Julich) (Ω-m)	0.526
ρ electrolyte (St. Gobain) (Ω-m)	2.116
ρ flow channels (Ω-m)	1.176 x 10^{-6}
k O2 electrode (W/m-K)	9.6
k O2 functional layer (W/m-K)	9.6
k H2 electrode (W/m-K)	13.1
k H2 functional layer (W/m-K)	13.1
k electrolyte (W/m-K)	2.16
k O2-flow channel (W/m-K)	27.0
k H2-flow channel (W/m-K)	72.0
Activation polarization	
$i_{0,ref}$ H2/H2O (A/m^2)	10000
$i_{0,ref}$ O2 (A/m^2)	1e8
Diffusion polarization	
ϕ electrodes (%)	35
ϕ functional layers (%)	25
ϕ flow channels (%)	87
L_t O2 electrode	2.00
L_t O2 functional layer	3.00
L_t H2 electrode	2.63
L_t H2 functional layer	3.33
K O2 electrode (m^2) (isotropic)	1.70 x 10^{-15}
K O2 functional layer (m^2) (isotropic)	1.00 x 10^{-15}
K H2 electrode (m^2) (isotropic)	8.01 x 10^{-15}
K H2 functional layer (m^2) (isotropic)	2.97 x 10^{-15}
K flow channels (m^2)	
In the flow direction	2 x 10^{-4}
Perpendicular to the flow direction	2 x 10^{-5}
Dimensions (width x length x thickness)	
H2 current collector / flow channel (mm)	40 x 40 x 1.000
H2/O2 inlet and outlet flow channels (mm)	5 x 40 x 1.000
H2 electrode (mm)	40 x 40 x 1.402
H2 functional layer (mm)	40 x 40 x 0.008
Electrolyte (mm)	40 x 40 x 0.010
O2 electrode (mm)	40 x 40 x 0.080
O2 functional layer (mm)	40 x 40 x 0.010
O2 current collector / flow channel (mm)	40 x 40 x 1.000

Table 2. Flow Rates and Species Concentration Input.

Julich Model

- Mass flow rate H2 side = 1.701e-6 kg/s
- Mass flow rate O2 side = 1.975e-5 kg/s
- Mass fractions H2 side : H2O = 0.217, H2 = 0.713, N2 = 0.00
- Mass fractions O2 side : O2 = 0.2329, N2 = 0.7671
- Operating pressure = 101.325 kPa
- Inlet temperature H2 and O2 side = 1077 K

St. Gobain Model

- Mass flow rate H2 side = 1.749e-5 kg/s
- Mass flow rate O2 side = 2.154e-5 kg/s
- Mass fractions H2 side : H2O = 0.36168, H2 = 0.04285, N2 = 0.59547
- Mass fractions O2 side : O2 = 0.2329, N2 = 0.7671
- Operating pressure = 85.0 kPa
- Inlet temperature H2 and O2 side = 1173 K

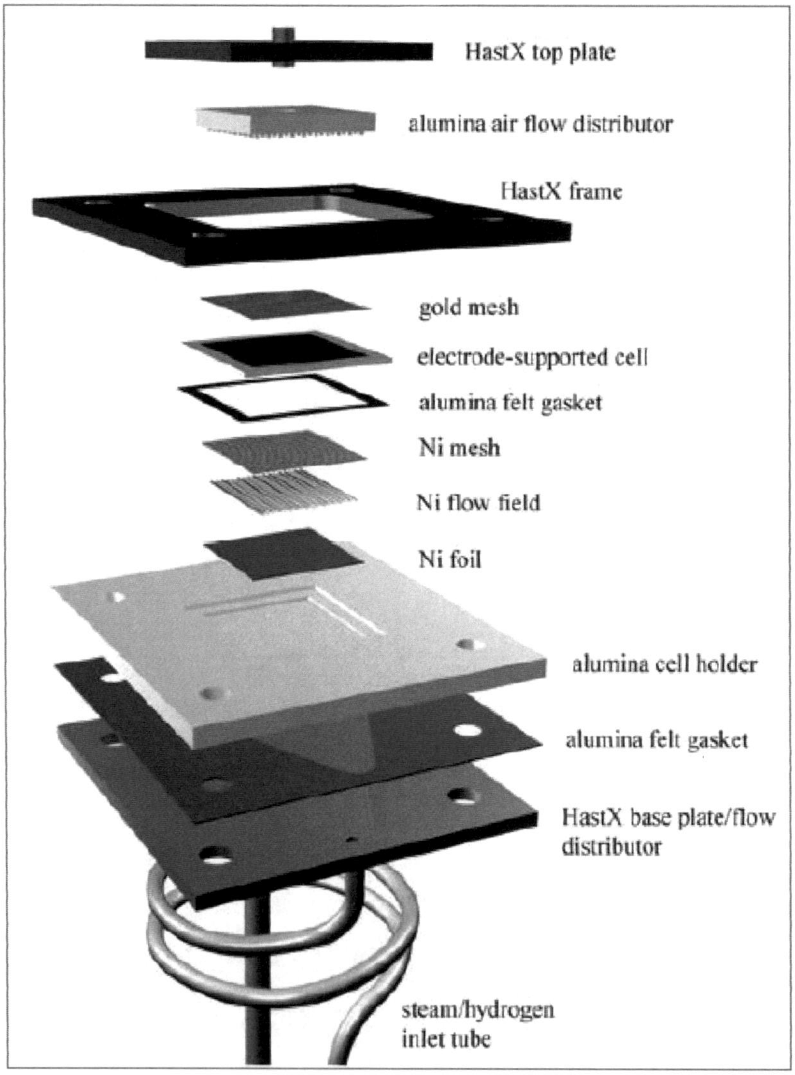

Figure 1. Detail of SOEC stack.

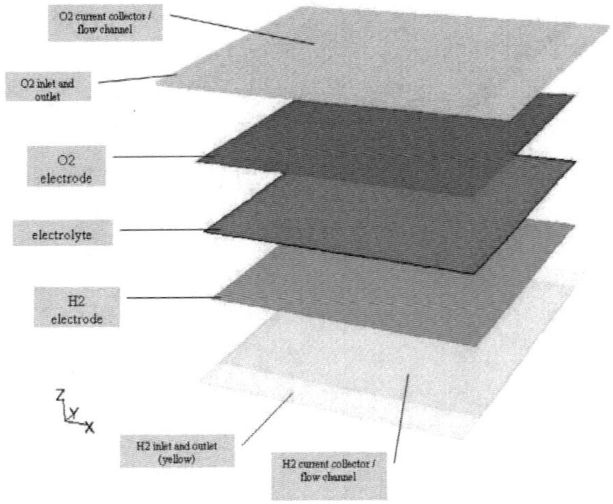

Figure 2. FLUENT single cell SOEC model.

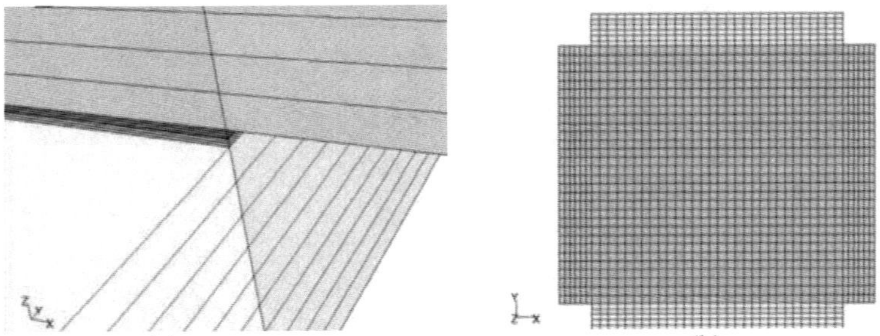

(a) (b)

Figure 3. Details of 3D numerical mesh; (a) close-up of corner showing vertical element stacking, (b) top view showing grid.

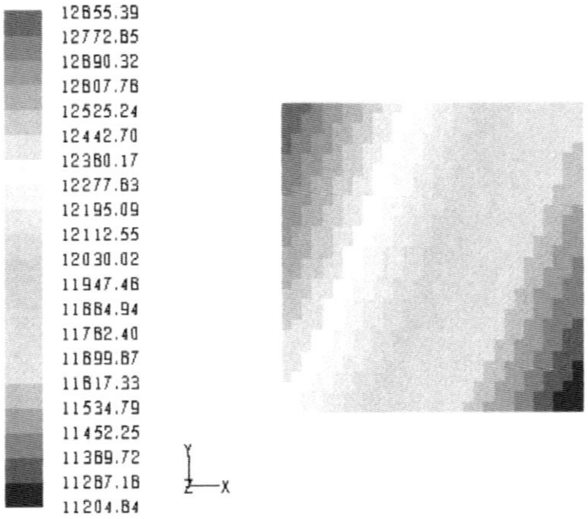

Figure 4. Contours of current density (A/m^2) at 19.2 A for Julich cell.

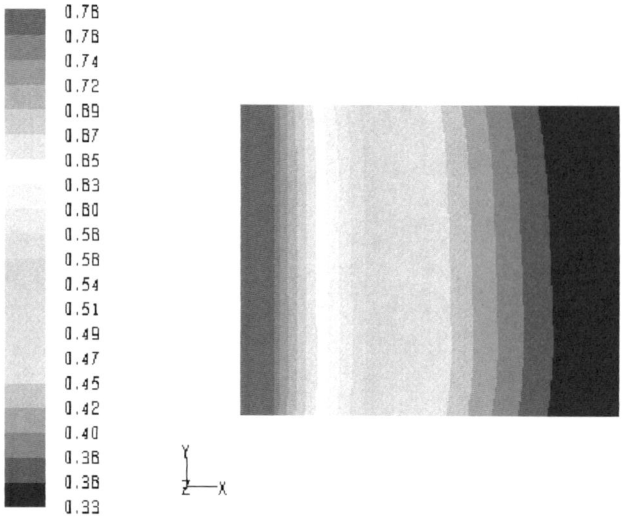

Figure 5. Contours of mass fraction of H2 for 19.2 A for Julich cell.

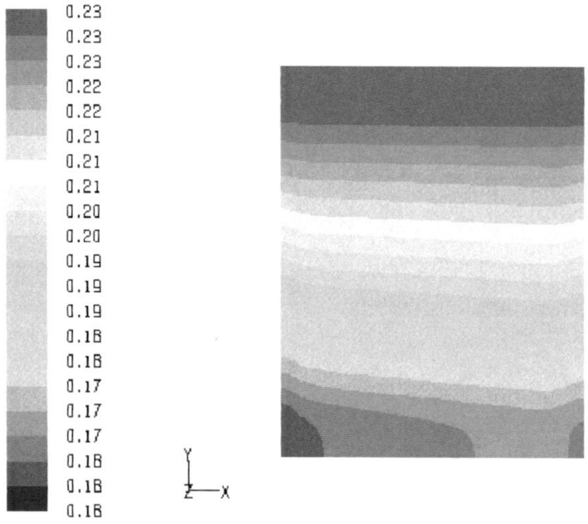

Figure 6. Contours of mass fraction of O2 for 19.2 A for Julich cell.

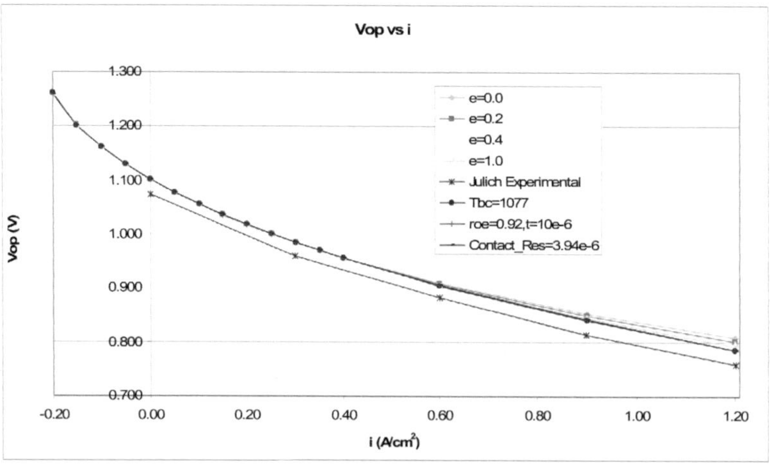

Figure 7. Vop versus current density for Julich cell in electrolysis and fuel cell mode for various parameters.

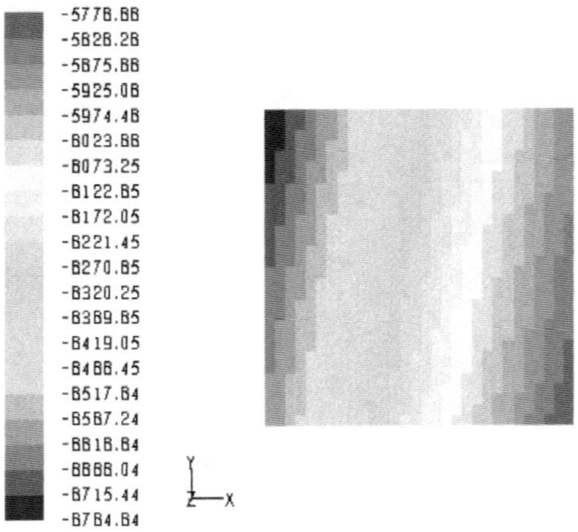

Figure 8. Contour of current density in electrolyte for St. Gobain cell at -10.0 A..

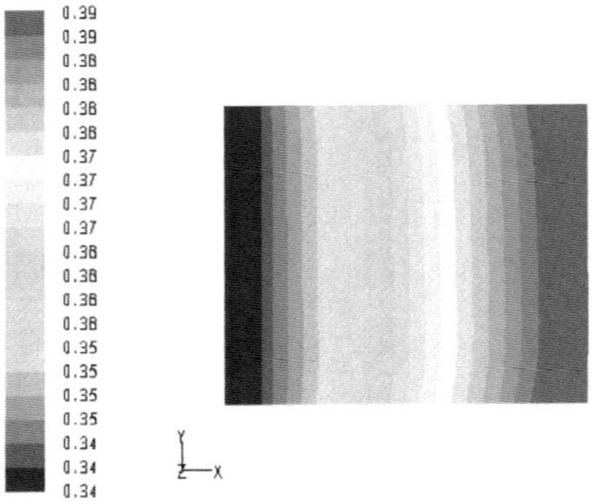

Figure 9. Contours H2 mole fraction for St. Gobain cell at -10.0 A..

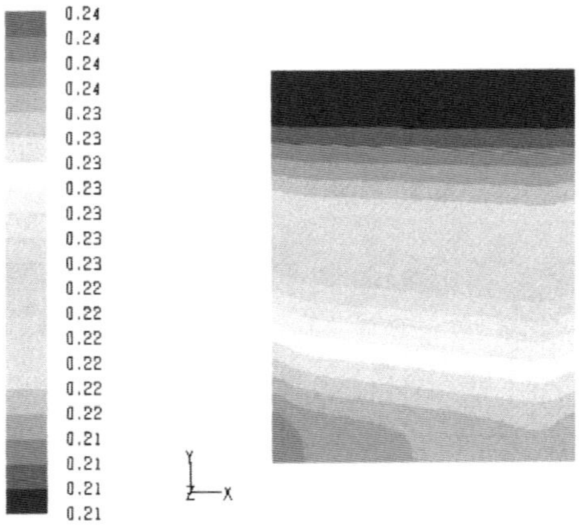

Figure 10. Contours O2 mole fraction for St. Gobain cell at -10.0 A.

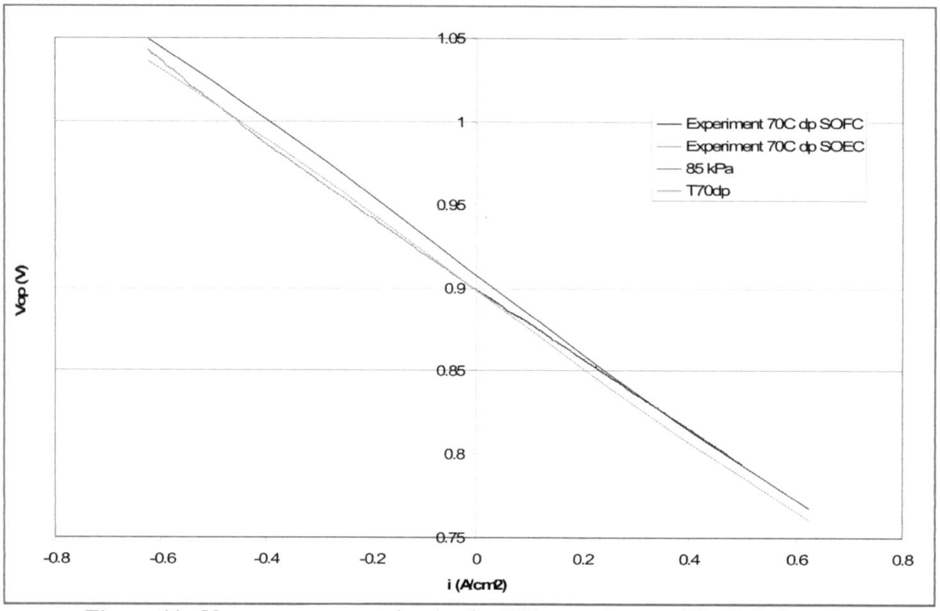

Figure 11. Vop versus current density for INL experiment of St. Gobain cell..

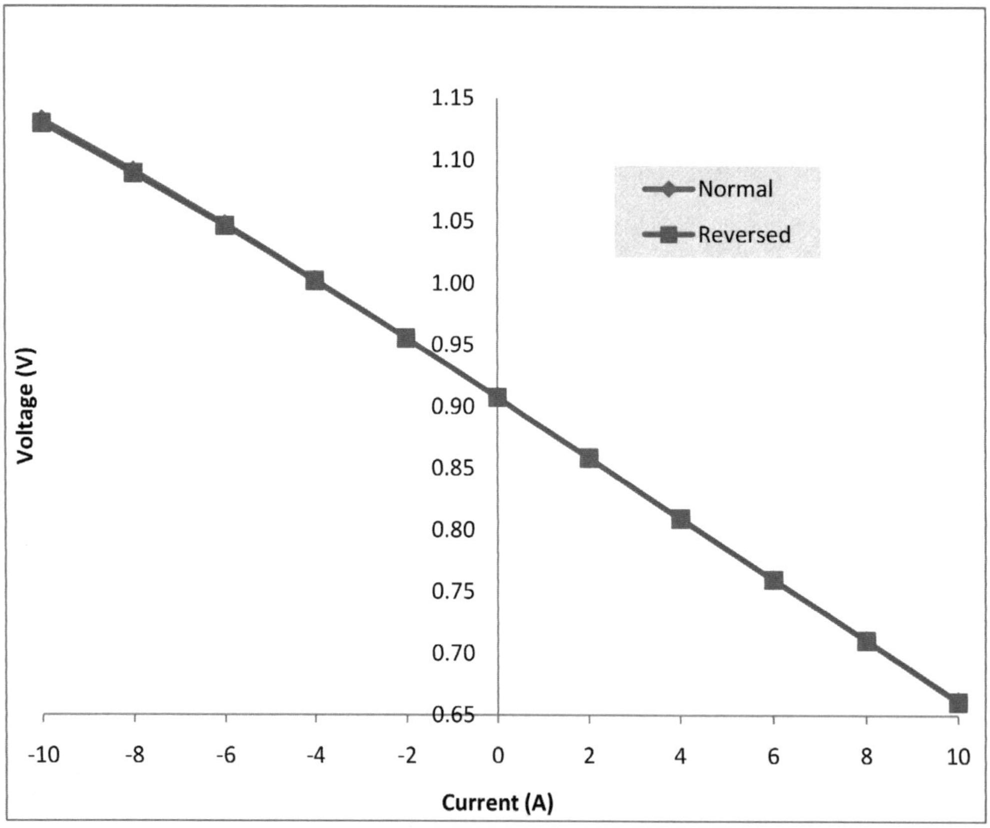

Figure 12. Vop versus current density for thick hydrogen and thick oxygen electrodes.

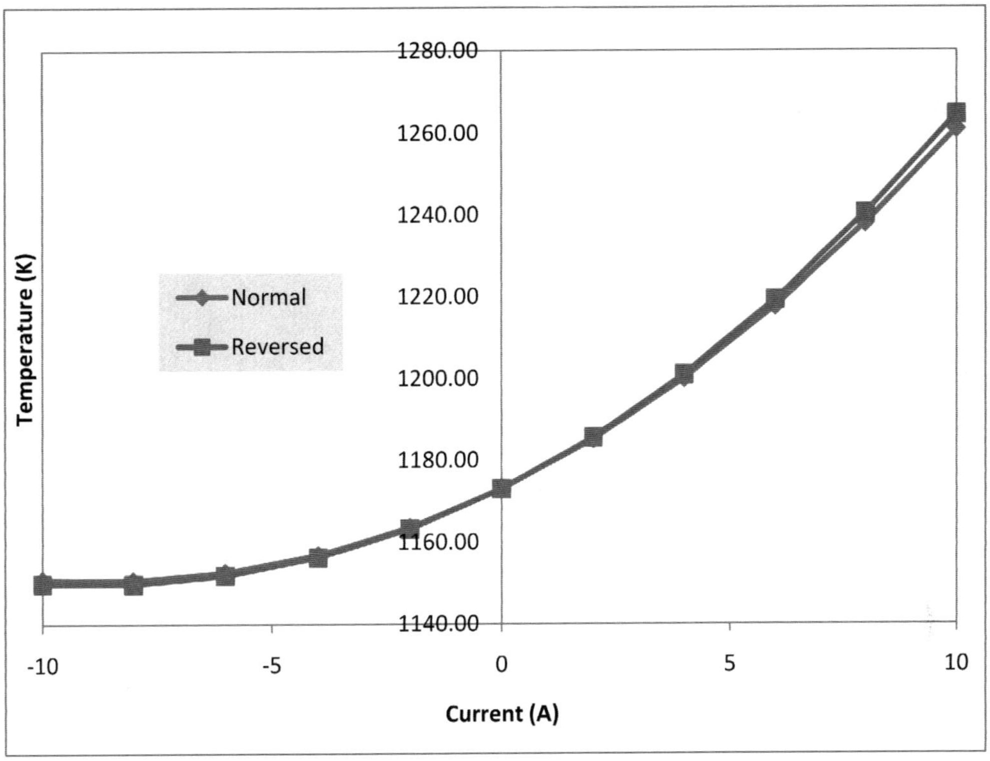

Figure 13. Temperature versus current density for thick hydrogen and thick oxygen electrodes.

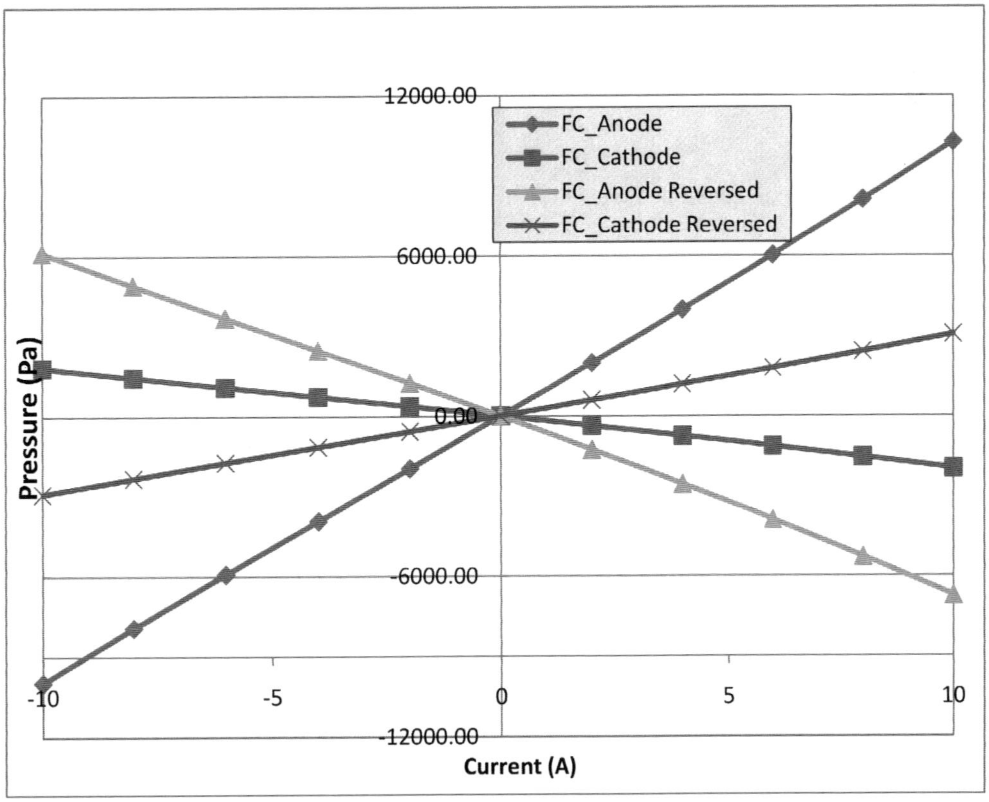

Figure 14. Functional layer pressure versus current density for thick hydrogen and thick oxygen electrodes.

AUTHOR INDEX

AUTHOR INDEX

AUTHOR INDEX

American Institute of Chemical Engineers
3 Park Avenue
New York, New York 10016-5991